OFFICE OF POPULATION CENSUSES AND SURVEYS

Visiting museums

A report of a survey of visitors to the
Victoria and Albert, Science and National Railway
Museums for the Office of Arts and Libraries

Patrick Heady

SS 1147

London: Her Majesty's Stationery Office

ISBN 0 11 691096 8

Acknowledgements

This survey could never have been carried out without the co-operation of a wide range of people. It is not possible to mention everyone who helped by name but particular thanks are due to our main contacts at each museum: Mr Physick and Ms Runyard at the Victoria and Albert Museum, Dr Robinson at the Science Museum and Dr Coiley at the National Railway Museum. I should like to thank them and their professional colleagues at the museums and at the Office of Arts and Libraries for their invaluable help in planning the survey and for their comments on drafts of the report. I should also like to thank the uniformed warders whose work, particularly at the museum entrances, was not made any easier by the presence of teams of interviewers.

Dr Alt, then of the Natural History Museum, was very generous with his experience of museum surveys. Thanks are due to Mobil for the use of a plan from one of their Victoria and Albert Museum broadsheets.

I should also like to thank all the OPCS staff who worked on the survey—both interviewers and the headquarters staff who helped plan the work and who processed the results. Particular thanks are due to Mrs Jones for her work on the interviewing procedures.

Finally, of course, I am very grateful to the members of the public who agreed to be interviewed, and also to their companions who had to wait while the interviews took place.

Needless to say, the shortcomings of this report are my own responsibility.

Contents

List of tables, figures and plans

1 Introduction

This report sets out the results of a survey carried out at three museums—the Victoria and Albert Museum, the Science Museum and the National Railway Museum—by staff of the Social Survey Division of the Office of Population Censuses and Surveys. The survey was commissioned by the Office of Arts and Libraries on behalf of the managements of the three museums to be studied. Its primary purpose was to provide information that would be of value to staff responsible for various aspects of the design and organisation of the three museums. The data was collected by means of interviews with members of the public who visited the three museums during the survey year, which ran from December 1979 to December 1980.

a. The museums

The Science Museum and the Victoria and Albert Museum, which occupy neighbouring sites in South Kensington, are descendants of a single South Kensington Museum whose history goes back to the Great Exhibition of 1851. However since 1909 they have been separate, indeed contrasting, institutions. The Victoria and Albert Museum is a museum of art and design exhibiting works from Britain, Europe and the East ranging in date from the middle ages to the present. The Science Museum deals with science and technology from earliest times up until today.

The plans of the two museums (see pages 2–5) show that both are housed in large buildings containing many separate galleries, that is to say areas of wall and floor space devoted to a particular topic or set of exhibits. The Science Museum's building is fairly plain, with a similar floor plan on every storey from the ground to the third floor. The Victoria and Albert Museum on the other hand occupies a rather complex late Victorian structure built over several years.

At the time the survey was being carried out both museums had plans for expansion. Since the end of fieldwork on the survey the Science Museum has inaugurated the Wellcome Museum of the History of Medicine on the recently opened fourth and fifth floors. The Victoria and Albert Museum has opened a new entrance on Exhibition Road which gives access to the main body of the museum as well as to the new Boilerhouse Project and to exhibition space in the Henry Cole Wing which has been converted for the purpose. The Tudor Art gallery, one of those studied in depth in Chapters 4 to 7 of the report, has been thoroughly renovated since the time the survey was carried out. The plans of both museums given in this report refer to the situation at the time interviewing took place and show the galleries that were open at that time—that is, December 1979 to December 1980.

The third museum covered by the survey was the National Railway Museum in York. This museum, which is linked administratively to the Science Museum, opened in 1975 bringing together exhibits from two previous museums in Clapham and York. It is housed in a building which, before substantial alterations, used to be a steam engine shed and the greater part of its display space is occupied by locomotives and rolling-stock ranged round two turntables on the floor of the main hall.

b. The scope of the survey

The study had three major objectives. The first was to provide an accurate description of each museum's visiting public in terms of age, sex, educational level and so on and to discover the proportions of visitors coming to each museum alone, with family or friends or in organised parties of different kinds.

The second objective was to provide information about people's visits to the museum: their reasons for visiting the museum in question, their expectations, the amount of time they spent in the museum and how much of it they attempted to see in one visit, which galleries they visited and how interesting they found each. Focusing in more detail on individual galleries, this aspect of the enquiry would attempt to discover which exhibits were looked at by most visitors and the points visitors particularly noticed about various exhibits as well as investigating the use made of labels, and other information sources such as telephone and loudspeaker recordings and television films.

In meeting these two objectives the survey would provide much information that was useful to the managements of the three museums. It would identify sections of the public for whom the museums had a particularly strong appeal as well as those whom they did not seem to be attracting so successfully. It would show which galleries had been particularly successful with different categories of visitor and provide information about visitors' interests and patterns of attention that might help to explain the appeal of different galleries. Information about the routes visitors took round each museum might help to explain why some parts of each museum were usually crowded while others were generally nearly empty.

However all this would provide a rather flat and mechanistic model of what happened in the three museums

VICTORIA AND ALBERT MUSEUM PLAN

LOWER FLOORS

KEY

- ▨ UPPER GROUND FLOOR
- ☐ GROUND FLOOR (STREET LEVEL)
- ▨ LOWER GROUND FLOOR

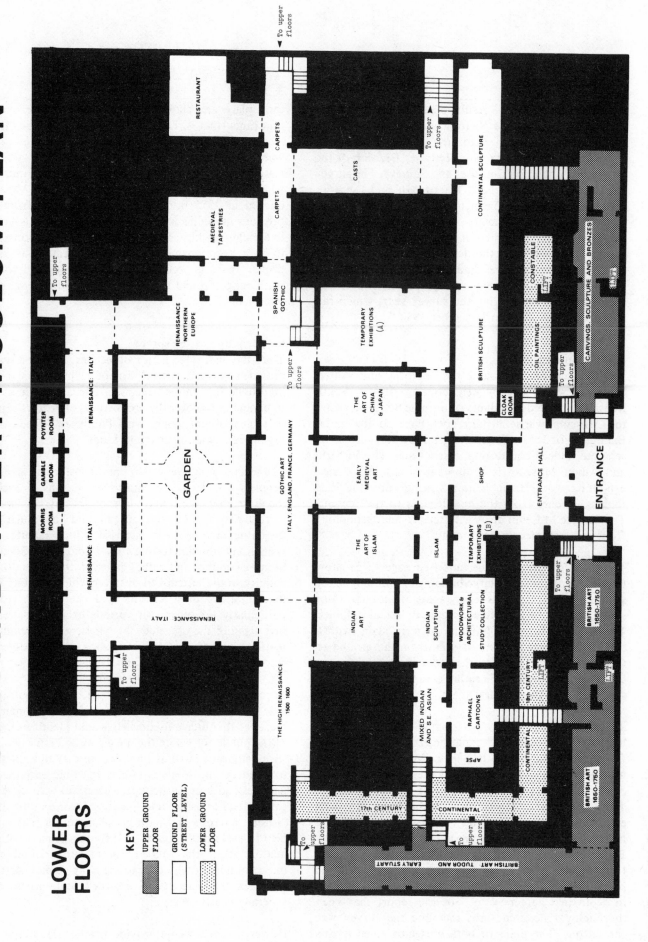

UPPER FLOORS

KEY

SECOND FLOOR
UPPER FIRST FLOOR
FIRST FLOOR

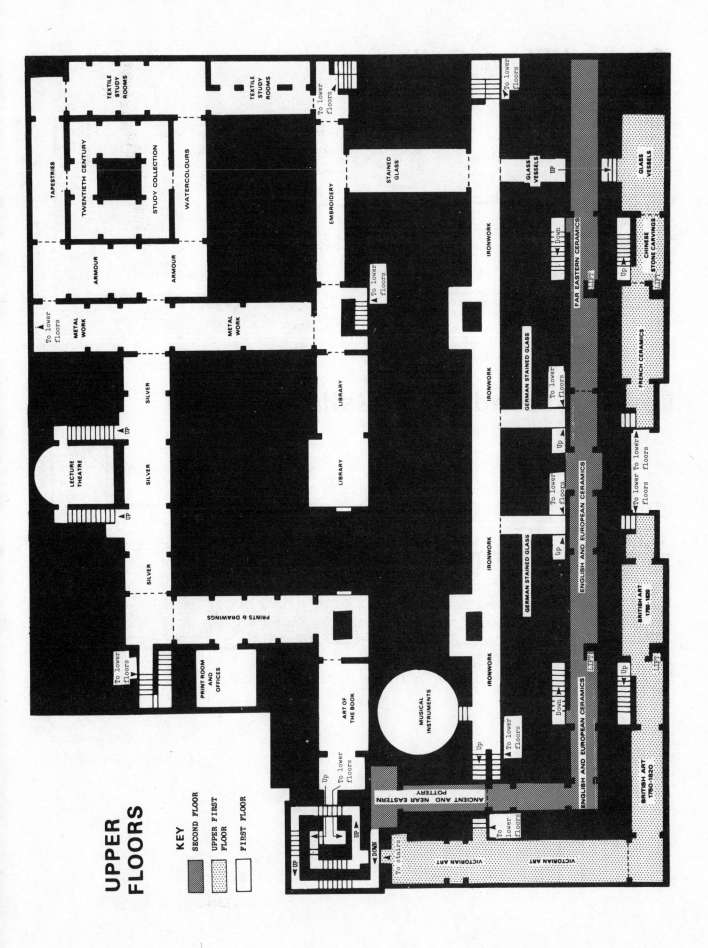

TEXTILE STUDY ROOMS

TAPESTRIES

TWENTIETH CENTURY

STUDY COLLECTION

WATERCOLOURS

TEXTILE STUDY ROOMS

To lower floors

ARMOUR

ARMOUR

METAL WORK

METAL WORK

EMBROIDERY

STAINED GLASS

To lower floors

IRONWORK

GLASS VESSELS

UP

To lower floors

GLASS VESSELS

LECTURE THEATRE

UP

UP

SILVER

SILVER

SILVER

SILVER

LIBRARY

LIBRARY

IRONWORK

IRONWORK

GERMAN STAINED GLASS

Down

UP

To lower floors

FAR EASTERN CERAMICS

LIFT

UP

CHINESE STONE CARVINGS

LIFT

PRINTS & DRAWINGS

GERMAN STAINED GLASS

UP

To lower floors

UP

ENGLISH AND EUROPEAN CERAMICS

FRENCH CERAMICS

To lower floors

To lower floors

LIFT

To lower floors

PRINT ROOM AND OFFICES

ART OF THE BOOK

MUSICAL INSTRUMENTS

IRONWORK

UP

To lower floors

ENGLISH AND EUROPEAN CERAMICS

Down

UP

BRITISH ART 1750-1820

LIFT

UP

LIFT

BRITISH ART 1750-1820

Up

To lower floors

Up

UP

ANCIENT AND NEAR EASTERN POTTERY

DOWN

To lower floors

To Stairs

VICTORIAN ART

VICTORIAN ART

VICTORIAN ART

SCIENCE MUSEUM PLAN

Ground Floor Galleries

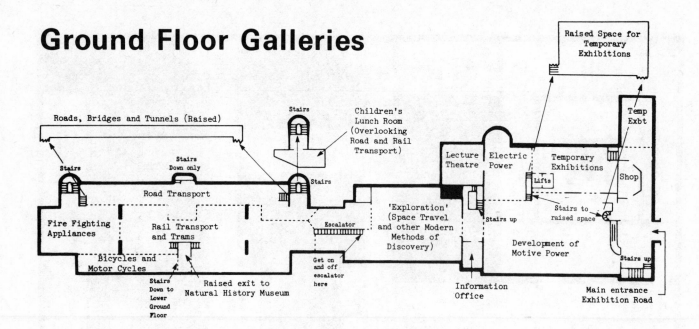

Lower Ground Floor Galleries

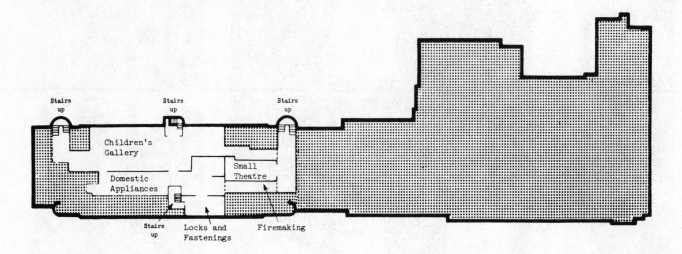

Third Floor Galleries

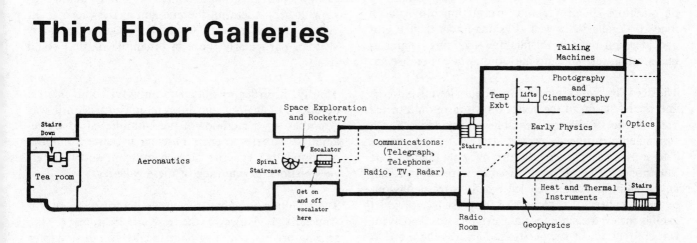

Second Floor Galleries

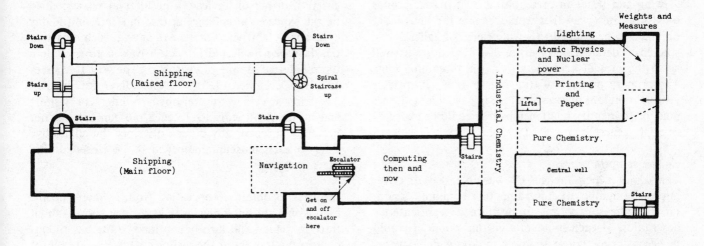

First Floor Galleries

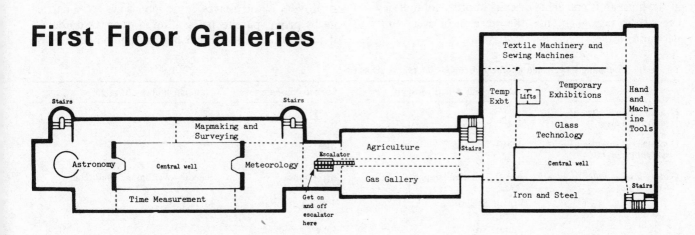

unless it could be related to visitors' own views about how well they thought the museums provided for their interests and requirements. The survey's third objective, therefore, was to ask visitors for their views on questions of display and information provision in individual galleries as well as asking about their use of and ideas on the general facilities—lighting, signposting, seating, refreshments and so on—at each museum.

The need for usable information about viewing patterns and their relation to the visitors' own tastes and to the presentation of different exhibits and galleries had a fundamental influence on the strategy of the survey. The museums, particularly the Victoria and Albert Museum and the Science Museum, were so vast that any attempt to obtain detailed reactions to the lay-out of each gallery, let alone the way each individual exhibit (or even each important exhibit) was displayed, was entirely out of the question. This was the more so since alterations in the organisation of the museums and in the contents and design of individual galleries would rapidly render much of this particularistic information out of date.

These practical considerations led to the decision that the survey should attempt to describe the styles of viewing and tastes in presentation of different groups of visitors in a way that would enable the knowledge obtained to be applied to a wide range of galleries not studied in detail themselves, and to the future as well as to the present organisation of the three museums. The primary purpose of studying viewing patterns, or visitors' opinions on presentation, in particular galleries must be to identify relationships that could be expected to apply generally.

c. The samples

The topics which the survey was intended to cover imposed a number of constraints on the sample design. First, if the survey was to produce a representative description of each museum's visiting public it would be necessary to choose a sample of days which covered different days of the week and was spread over a whole year in order to avoid seasonal effects. On each of the days selected interviewers would have to be positioned at the museum's entrances or exits in order to interview every 'n'th person to enter the museum or every 'n'th person to leave.

Secondly, the idea that the survey should obtain accounts of which parts of the museum the visitors had looked at, and the visitors' views on the general facilities provided, implied interviews of 20 minutes or so with people leaving the museum at the end of their visits. Twenty minutes would be longer than many visitors, particularly those in organised parties, could spare.

Thirdly, in order to obtain meaningful comments on questions of lay-out and information provision it was desirable, since methods of presentation varied widely amongst galleries at each museum, to select a number of contrasting galleries at each museum and to interview visitors leaving each of these galleries.

The resulting sampling scheme is described in full in Appendix 1. However the essential features of the scheme are set out diagrammatically for this chapter in Figures 1.1 and 1.2. Three types of sample were taken at each museum.

i. Count-based sample of museum leavers

For the first type of sample, the count-based samples of museum leavers, people were counted as they left the museums and an attempt was made to interview every 'n'th person to leave (the interval set depended on the expected attendance at that museum on the day concerned). This was done on 16 days at each museum, the days being selected in such a way as to provide a representative sample of visitors for the year December 1979 to December 1980 (see Appendix 1 for details). The interviewers were stationed to intercept visitors leaving the main entrance to each museum. It was not necessary to cover any other exits since the alternative exits to each museum happened to be closed during the survey year.

In order to obtain co-operation from as many people as possible the interview was kept very short—about three minutes—and largely confined to the basic items of information about the visitor's background, who if anyone was with him, and his reasons for visiting the museum, as shown in Figure 1.2. In order to obtain information on visitors who would not have been able to provide it themselves some interviews were carried out by proxy, parents being allowed to speak on behalf of small children and English speaking companions on

Figure 1.1 Sampling scheme and numbers of eligible interviews achieved

	Victoria and Albert Museum	Science Museum	National Railway Museum
Count-based samples of museum leavers	891	787	920
Quota-based samples of museum leavers	1,003	1,045	596
Quota-based samples of gallery leavers	Art of China and Japan 232	Aeronautics 255	Balcony overlooking Main Hall 248
	British Sculpture 240	Printing and Paper 222	Main Hall 263
	Tudor Art 251	Time Measurement 227	
	Continental 17th Century Art 236	Exploration 243	

Figure 1.2 Sampling scheme: information sought and criteria for eligibility

	Count-based samples of museum leavers	Quota-based samples of museum leavers	Quota-based samples of gallery leavers
Information sought	social characteristics of the visitor	social characteristics of the visitor	social characteristics of the visitor
	companions (if any)	companions (if any)	companions (if any)
	reasons for visiting the museum	reasons for visiting the museum	exhibits looked at in the gallery
		galleries looked at during the visit, and how interesting the visitor found them	points recalled about an individual exhibit
		use of, and views on, the general facilities provided	views on display and information provision
			how interesting the visitor found the gallery
Who was eligible for the sample	all visitors	visitors aged 11 and over not with organised parties	visitors aged 11 and over, including children with school parties but excluding other organised party visitors

behalf of foreign visitors whose command of English was inadequate for the interview.

Figure 1.1 shows that the number of valid interviews obtained from the count-based samples varied from just under 800 at the Science Museum to a little over 900 at the National Railway Museum. Not everyone designated by the count provided an interview. The proportion of designations that resulted in a valid interview was 74 per cent at the Victoria and Albert Museum, 67 per cent at the Science Museum, and 83 per cent at the National Railway Museum. The remaining designations were accounted for first by people who were ineligible for interview—eight per cent at the Victoria and Albert Museum, eight per cent at the Science Museum and one per cent at the National Railway Museum. This category covered the museums' own staff, delivery and maintenance staff, people who turned back into the museum at the last moment, and individuals who had already left the museum temporarily during the day and had thus already had a chance of selection. Secondly 12 per cent of the individuals designated by the count at the Victoria and Albert Museum refused to be interviewed, as did 13 per cent at the Science Museum and six per cent at the National Railway Museum. Finally the interviewers were unable to contact some members of the designated samples. This accounted for the remaining six per cent at the Victoria and Albert Museum, 11 per cent at the Science Museum and ten per cent of the designated sample at the National Railway Museum. Many members of this last category were people who left the museums at the end of the day after the interviewers had been obliged to stop interviewing in order not to interfere with the process of closing the museum. The pattern of 'non-response', that is of refusals and 'non-contacts', is discussed further in Appendix 1.

Some practical considerations meant that the people visiting the museums on different days had different overall chances of selection for the count-based samples. In order to ensure that certain times of year, for instance Christmas and Easter holidays, received at least some representation, days at those times of year were given a higher than usual probability of selection. The other major consideration was the sub-

stantial variation over the year in the daily attendance at each museum. In order to provide a reasonable flow of work for the interviewers the sampling interval used in connection with the count of people leaving the museum was adjusted in the light of the level of attendance expected on the day concerned. For instance the longest sampling interval used at the Science Museum was one in three hundred. By contrast interviewers were instructed to contact every fortieth person to leave on two days which were expected to produce a sparse attendance.

In order to prevent these differences in selection probabilities biasing the information obtained from the count-based samples the results have been adjusted ('weighted') to give more weight to answers from people interviewed on days when the overall chances of selection were low. In addition the data has been further weighted to compensate for low response rates obtained from children aged ten or under visiting in organised parties. The purpose of weighting the data in these ways is to make it fully representative of the public who visited each museum during the survey year. Full details of the system of weighting used are given in Appendix 1.

ii. Quota-based samples
In the second type of sample, quota-based samples of museum leavers provided detailed information about their visits which could not have been collected in the very short interviews which members of the count-based samples were asked to give. The essential idea of quota sampling is that interviewers are asked to interview given numbers of individuals in certain pre-specified categories. In this case the categories were defined in terms of age, sex and whether the person concerned was visiting the museum alone or with others. No attempt was made to interview children aged ten or less (at the Science Museum particularly this meant excluding a substantial portion of the visiting public—see Table 2.3), people who did not speak English or people visiting the museum as members of organised parties. Just over a thousand quota sampled interviews were obtained with people leaving each of the two South Kensington museums. Five hundred and

ninety-six interviews were obtained with people leaving the National Railway Museum.

The third type of sample was of people leaving certain specified galleries in each museum. The galleries were chosen in consultation with staff of the museums in such a way as to provide a range of different types of subject matter and styles of presentation. Visitors were asked to recall details of their visit to the gallery in question and to comment on its presentation. The samples were quota-based using the same criteria as the quota-based samples of museum leavers, except that in the galleries interviewers were asked to obtain some interviews with children, aged 11 or over, visiting the museums with school parties. At both the Victoria and Albert Museum and the Science Museum interviews were conducted with visitors leaving each of four different galleries. At the National Railway Museum only two galleries (one of them the Main Hall) were selected. As Figure 1.1 shows the number of interviews obtained ranged from 222 in the Printing and Paper gallery at the Science Museum to 263 in the Main Hall at the National Railway Museum.

The quota samples are not fully representative of each museum's visiting public in the same way that the count-based samples are. Indeed the composition of the quota samples differs from that of the count-based samples with respect to age, sex, the numbers of solitary visitors and some other factors. This divergence from the count-based samples is acceptable because the primary purpose of the quota samples was not to make estimates of the proportions of visitors with particular characteristics but rather to explore the relationships between different factors, for instance to find out whether visitors' educational backgrounds are related to the styles of presentation they prefer.

No attempt will be made to adjust the quota sample data to make it precisely comparable with the results obtained from the count-based samples. Instead, when data from quota samples does have to be used to make estimates of proportions (for instance the proportions of visitors satisfied with various of the museum's facilities) care will be taken to point out ways in which the composition of quota samples are likely to have biased the results.

Appendix 1 discusses the relationship between the count-based samples and the quota samples.

iii. Sampling variability
The results of this report are of course affected by sampling variability. This affects the results from both the count-based and quota-based samples. The problem for the reader is how to decide which estimates are reliable and which are less certain. Formal confidence intervals and tests of significance have not been used in this report. However the text of the report does provide guidance. The more reliable results have been reported in definite language while those which depend on more variable figures have been described in a tentative way. The more reliable results are those based on larger numbers. Where a contrast between different museums is drawn it is more reliable, other things being equal, if the percentage difference in the factor of interest is large. Relationship between visitors' characteristics and their viewing patterns and reactions are most reliable if they are based on comparable figures from several different galleries or from all three museums, or if the results derived from the gallery interviews are corroborated by data from the samples of museum leavers.

d. The plan of this report
The following chapters will deal with topics in roughly the same order as they would arise during a museum visit. Chapter 2 will describe the public who visit each museum. Chapter 3 will analyse the reasons they give for their visits and attempt to discover which parts of each museum's collection play particularly important roles in drawing visitors to the museum.

Chapters 4 to 7, based on the gallery samples, deal with visitors' reactions in particular galleries—showing which exhibits they look at, what strikes them about specific objects and discussing their views on display and information provision. Chapter 8 looks at the overall pattern of visits to each museum—how many galleries people look at during a visit, which galleries they look at and which they walk through, and how often they visit the museum. Chapter 9, which is based on the interest levels reported for each gallery they stopped to look at by people leaving the museums, will investigate whether the results derived in Chapters 4 to 7 can be generalised to explain interest levels in the full range of galleries at each museum.

Chapter 10 reports on visitors' use of facilities such as the museums' shops and restaurants and on how satisfied they are with various aspects of the running of each museum. Chapter 11 focuses specifically on certain categories of visitor: children aged ten or less, children visiting with school parties, and visitors from overseas. Chapter 12 rounds off the story by discussing the level of satisfaction with their visit as a whole reported by people leaving each museum and relating overall satisfaction with the visit to some of the factors discussed in more detail in previous chapters.

This report will deal with the results for all three museums, different as they are, in the expectation that the differences and similarities that emerge will shed more light on the patterns of visiting at each museum than would three separate reports. Even so within each chapter most of the results will be presented separately for each museum so that readers who are concerned specifically with one of the museums should have little difficulty in locating the material that is relevant for them.

e. Outline of some of the main findings
Chapter 2 shows that the museums differ in the visitors they attract. Over 60 per cent of visitors to the two

technical museums were men or boys, compared to only 43 per cent at the Victoria and Albert Museum. The Science Museum attracts the youngest public: 55 per cent of its visitors are aged 20 or less compared to 29 per cent at the Victoria and Albert Museum and 32 per cent at the National Railway Museum. The museums draw their British visitors predominantly from their own part of the country. Thus 79 per cent of the Victoria and Albert Museum's UK resident visitors and 67 per cent of those at the Science Museum live in London and the South East—compared to only ten per cent of the National Railway Museum's British visitors.

Sixty-four per cent of the National Railway Museum's visitors were viewing it for the first time while just under half the people interviewed at the two London museums were first-time visitors. The National Railway Museum was the museum most visited in family groups: if people visiting with organised parties are excluded 80 per cent of its remaining visitors were visiting with other members of their families. The comparable figures for the Science Museum and the Victoria and Albert Museum were 60 per cent and 43 per cent respectively.

The reasons given for visiting each museum are discussed in Chapter 3. As would be expected many visitors at all three museums said that they had come out of general interest, because of the museum's reputation or because of a specific interest in its contents. Special temporary exhibitions attract 26 per cent of the Victoria and Albert Museum's visitors. Many people visit museums altruistically in order to accompany someone else. People visiting in order to accompany others form 11 per cent of the public at the Victoria and Albert Museum, 33 per cent at the Science Museum and 40 per cent at the National Railway Museum. Seven per cent of the Victoria and Albert Museum's visitors, and rather fewer at the other two museums gave a reason for visiting that was connected with their work or studies.

As well as special temporary exhibitions some parts of each London museum's permanent collections were particularly big 'draws'. At the Victoria and Albert Museum these were costumes (though the Costume gallery was closed for renovation work at the time) paintings and furniture. At the Science Museum five of the subjects covered by the permanent collections could be said to constitute 'star attractions'. These subjects were road and rail transport, aircraft, the Children's gallery, space travel and computing.

Chapter 4 demonstrates definite differences in visitors' interest levels between the ten galleries where interviews took place within the museums. It also shows that individual visitors' ratings of the galleries for enjoyment correspond very closely to their assessments of how interesting they are. When visitors talk about enjoying looking at a gallery they mean that it has captured their interest.

The visitor's prior interest in the gallery's subject matter plays an important role. Results set out in Chapter 5 show that visitors who had enough initial interest to seek out a particular gallery were generally much more likely to find it interesting than those who merely came across it while walking through the museum.

Visitors do not generally work their way systematically through all the contents of those galleries they decide to look at, choosing instead to stop and look at only some of the items on show. There is no simple formula giving a single set of qualities that an interesting exhibit must have. Indeed Chapter 5 gives varying 'comment profiles' of exhibits which attracted particular attention—based on what our informants were able to tell the interviewers about the exhibits concerned.

Chapter 6 starts by investigating the rather complex connection between attractive presentation and visitors' interest. The connection works both ways: if a visitor finds that a gallery is attractively presented this does appear to enhance his or her interest. On the other hand if the visitor has a prior interest in the subject of the gallery he is more likely to consider that it is attractive. Among the features which visitors found attractive were spaciousness, adequate lighting and a variety of 'special effects' ranging from period room settings at the Victoria and Albert Museum to aircraft suspended from the ceiling in the Aeronautics gallery at the Science Museum. Also important was a rather indefinable quality best described as the absence of dreariness.

It was important that the information provided in a gallery should be presented in a way that visitors could understand. Information that was a bit too elementary did not appear to interfere with visitors' enjoyment but information that was not elementary enough did appear to limit people's enjoyment.

Some galleries at the Science and National Railway Museums used audio-visual techniques as well as printed labels and notices to convey information. The different media were valued by visitors for different reasons. Print was praised as a medium from which information could be extracted very quickly, for allowing visitors to go at their own pace and because visitors could pick out just the information that interested them. The audio-visual media were preferred by people who found them easier than print. A number of people seem to have enjoyed using their ears as well as their eyes.

Chapter 7 shows that print is relatively popular with older and with better educated people while audio-visual media were particularly popular with younger visitors and with visitors who had finished full-time education when relatively young. Age was strongly associated with the level of interest visitors reported for each gallery as a whole. The older the visitor the more likely he or she was to judge the gallery 'very interesting'. One attitudinal factor was associated with the level of interest visitors expressed. Visitors who thought that it was important to learn something from a museum visit tended to report higher interest levels than people who said they *"just enjoyed looking at things"*.

Chapter 8 shows that people who were visiting the museums on their own were more likely than other visitors to have been to the museum concerned before. First visits tend to be more wide-ranging—in terms of the numbers of galleries visitors stop to look at—than are return visits, particularly visits by people who have been to the museum concerned several times in the previous 12 months. However even first-time visitors usually only stop to look at exhibits in a comparatively small proportion of the two London museums' numerous galleries.

In the two London museums visitors' choices of which galleries to look at are influenced both by the appeal of the galleries' subject matter and by the structure of the museum building itself. Special temporary exhibitions attract some new visitors to each London museum but their main effect on attendance is that they provide a reason for return visits by people who already know the museums concerned.

Chapter 9 is based on visitors' interest ratings of all the permanent galleries they had stopped to look at. It provides some confirmation of earlier findings concerning the role of intrinsic interest in the galleries' subject matter and of different types of presentation. In particular, galleries devoted to four of the five 'star attractions' at the Science Museum receive the highest interest ratings.

At all three museums the great majority of visitors who had asked members of staff for information reported that their replies had been very helpful. Most of the questions asked by visitors to the two London museums related to help finding their way round the museum. So it is not surprising that Chapter 10 also records that many visitors to the Victoria and Albert Museum and the Science Museum thought improvements were needed in the signs provided to help people find their way round.

The first part of Chapter 11 is based on the replies of parents concerning the reactions of their under-eleven year old children to galleries in the Science Museum and the National Railway Museum. The galleries that appeal most to these youngsters are, if their parents are to be believed, the same ones that are popular with visitors in other age groups. The appeal of these galleries seems to cut across generation divisions.

Chapter 11 also discusses foreign visitors. Overseas visitors whose first language was not English constituted 14 per cent of the count-linked interviews at the Victoria and Albert Museum and 11 per cent of the interviews at the Science Museum—but only three per cent at the National Railway Museum in York.

Chapter 12 provides further evidence of the central importance of interest for visitors' enjoyment of museums. Visitors who had found most of the galleries they stopped to look at very interesting were much more likely than others to report that the visit as a whole had been very enjoyable. Visitors' views on the practical organisation of the museum did not appear to be strongly related to their level of enjoyment—though problems with the directional signs did seem to have spoiled the enjoyment of some.

Visitors were ready to put a good deal of effort into their visits. Seventy-six per cent of visitors who had spent three hours or more in the Science Museums reported that they felt fairly or very tired. (This figure is derived from the quota sample of museum leavers.) However there was no sign that tiredness had stopped them enjoying their visits.

f. Conventions used in tables

Percentages are rounded to the nearest whole number. This results in some table columns or rows adding to slightly over or slightly under 100 per cent. Percentages of less than 0.5 are rounded down to zero. A dash indicates that no figure was calculated for the table cell concerned (generally because the base number was unusually small).

It is a commonplace of survey research that questions are frequently not answered by all the sample members to whom they apply. The shortfall, which is generally very small—for example, one or two per cent—typically arises because a few informants were not able to answer the question concerned or because interviewers occasionally missed it out. In order to make the tables easier to read we have not shown the percentage of 'no answers' at each question. Instead the following convention has been adopted. The base numbers quoted are the numbers of individuals to whom the table applied. The percentages however have been calculated using the slightly smaller number of people who actually answered as the denominator. The base numbers used for sub-sets of the data are the number of visitors definitely known to belong to that sub-set. Because of 'no answers' these base numbers can add to slightly less than the total sample.

In the case of the count-based samples weighted base numbers have been given. These are proportional to the total weight assigned to the individuals whose answers have contributed to the part of the table to which the base number refers. The weighted base for the whole count-based sample at each museum has been set arbitrarily at 1,000—so that if part of a table was based on all the eligible interviews at, say, the Victoria and Albert Museum the weighted base would be given as 1,000. An approximate estimate of the number of individuals who contributed to any part of a count-based table can be obtained by dividing the weighted base by 1,000 and multiplying the result by the number of eligible count-based interviews achieved at the museum concerned (891 at the Victoria and Albert Museum, 787 at the Science Museum, and 920 at the National Railway Museum).

The weighted base numbers have been rounded to the nearest ten. As a result the base numbers for the rows or columns of a count-based table may add to either slightly less or slightly more than the base number for the table as a whole.

2 Each museum's visiting public

a. General

Perhaps the most natural question to ask about the museums' visitors is: What kinds of people are they? The answer, or rather answers, will set the context for the remainder of this report. Detailed information about the behaviour and views of any category of visitors would be of limited use without some idea of what proportion of the museum's visitors belong to that category. Beyond that, information about the kind of people who visit each museum may give a first glimpse of the sort of places the museums are for their visitors and how the museums may differ from each other.

The tables in this chapter will be derived from the weighted versions of the count-based samples of museum leavers. As was said in Chapter 1, these provide a representative picture of each museum's visiting public. Before describing the results themselves a word is needed about the way in which they are representative. Strictly speaking what the sample represents is visits, so that the statements *"Ten per cent of visitors belong to this category"* or *"A third of the visiting public gave that reason for their visit"* should be understood to mean that 'ten per cent of visits are made by people who belong to this category' or 'a third of all visits are made by people who give that reason'.

The reason for choosing this sample of visits was that it provided a representative picture of daily attendance at each museum during the year (giving greatest weight to the days on which attendance was highest). However in order to be representative in this sense the sample had, necessarily, to be unrepresentative in another. Since the sample is based on visits, individuals making 'n' visits during the survey year have 'n' chances of selection compared with only one chance for individuals who visited the museum only once during the survey year. As a result the sample is unrepresentative of the year's visitors to each museum in the sense that it gives less weight to people who visited only once than to people who visited the museum repeatedly.

The question of what distinguishes frequent visitors from people who visit the museums less often is, of course, an important one and will be discussed in Chapter 8 which deals with different patterns of museum visiting.

b. Who visits the three museums

Table 2.1 gives the proportions of male and female visitors at each museum. It shows that the Victoria

Table 2.1   Percentage of male and female visitors at each museum

	Victoria and Albert Museum	Science Museum	National Railway Museum
	%	%	%
Male	43	64	61
Female	57	36	39
Weighted base	*1,000*	*1,000*	*1,000*

Source of table: weighted results from count-based samples of museum leavers.

Table 2.2   Percentage of male visitors at each museum by age

Age (years)	Victoria and Albert Museum	Science Museum	National Railway Museum
Up to 10	49%	60%	46%
11–20	34%	67%	71%
21–30	48%	58%	63%
31–40	48%	61%	57%
41 and over	41%	68%	62%

The base for each figure (that is the proportion of the weighted sample belonging to that age range) is shown by the percentage given for that age range in Table 2.3.

Source of table: weighted results from count-based samples of museum leavers.

Table 2.3   Age distribution of visitors to each museum

Age (years)	Victoria and Albert Museum	Science Museum	National Railway Museum
	%	%	%
Up to 10	6	19	11
11–20	23	36	21
21–30	28	15	15
31–40	16	15	27
41–50	12	8	11
51–70	13	6	15
71 and over	3	0	1
Weighted base	*1,000*	*1,000*	*1,000*

Source of table: weighted results from count-based samples of museum leavers.

and Albert Museum is visited by women rather more than by men. This sets the Victoria and Albert Museum apart from the other two museums, over 60 per cent of whose visitors are male. Table 2.2 takes different age groups separately and indicates that the tendency of the Science Museum and the National Railway Museum to attract a higher proportion of male visitors than does the Victoria and Albert Museum holds for all age groups from 11 upwards. The Victoria and Albert Museum and the National Railway Museum both appear to be visited by approximately equal numbers of boys and girls aged ten or under, though even in this age range the Science Museum's public may have a masculine bias.

The age distributions of visitors at each museum, set out in Table 2.3 and Figure 2.1, produce a different contrast. In this case it is the Science Museum which stands out as the museum with the youngest visiting public. Over half of its visitors are under 21, compared with a third or less at the other two museums. The Science Museum has the highest proportion of visitors in the under 11 age group, almost twice as many proportionally as the National Railway Museum, which in turn has almost twice as many as the Victoria and Albert Museum. The 11-20s are the peak age group at the Science Museum, over a third of its visitors belonging to this age band. At the Victoria and Albert Museum the 21-30 age range is most prominent while at the National Railway Museum the distribution seems to reach a peak between the ages of 31-40. Beyond the age of 40 the distribution declines steadily at all three museums.

This last point is interesting since it shows that none of the three museums included in this study conforms to the commonly held view that museums appeal particularly to older people. In fact the proportion of

visitors aged over 50 at the Victoria and Albert Museum and the National Railway Museum is only about half what would be expected from the proportion of older people in the population generally, while at the Science Museum it is lower still. Travelling to museums may be difficult for the older members of the age group. Anticipating slightly we can speculate that one other reason for the decline might be that as people grow older and their children start to grow up they and their children stop going on outings together. Those visitors for whom a trip to a museum was primarily an excuse for a family outing would therefore cease to have a motive for visiting museums.

The 11-20 age range includes between 21 per cent and 36 per cent of each museum's visitors. It is of course far from being a homogeneous group, including quite young children as well as college students and people with full-time jobs. Table 2.4 shows how this group is made up at each of the three museums. There is a marked difference between the Victoria and Albert Museum, where 40 per cent of the category is made up of students at schools or colleges who have con-

Figure 2.1 Age distribution at each museum (from Table 2.3)

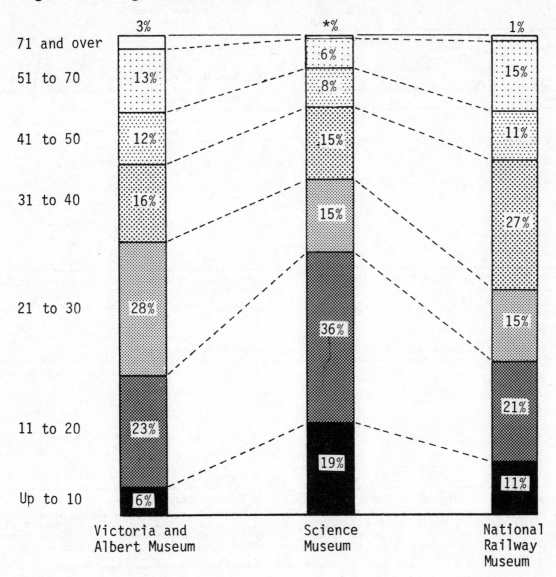

tinued their full-time education for more than a year beyond the minimum British school-leaving age, and the other two museums, where more than two thirds of the category are school children aged 11–16.

Table 2.5 deals with the major subject areas studied during the previous two years by the students who visited each museum. For this purpose we have classed as a student anyone aged 17 or over who is still in full-time education. The age of 17 was chosen since it was felt that by that age most students at secondary schools would have started to specialise, so that questions about which were the major subjects studied would be meaningful for them. In fact some of the young people who were counted as students had not yet specialised in any particular subject area; these students have been classed as pursuing 'general studies'.

The subjects have been divided into four main groups in Table 2.5 'Visual arts' include sculpture, fine art, design and art history. 'Other arts subjects' include history, literature, foreign languages, drama and music. 'Science' refers to biology, the physical sciences, mathematics and engineering. 'Other subjects' include social administrative and business studies, along with a wide range of vocational training ranging from medicine, through architecture to catering and domestic science. In Table 2.5 this somewhat miscellaneous collection of subjects has been grouped together with 'general studies'.

Table 2.4 Status of 11–20 years olds at each museum

	Victoria and Albert Museum	Science Museum	National Railway Museum
	%	%	%
School children aged 11–16	47	71	81
Aged 17–20 in full-time education	40	19	7
Finished full-time education	12	10	12
Weighted base	*230*	*360*	*210*

Source of table: weighted results from count-based samples of museum leavers.

Table 2.5 Main subjects studied during last two years by full-time students aged 17 or over

Main subjects	Victoria and Albert Museum	Science Museum	National Railway Museum
	%	%	%
Visual arts	42	8	—
Other 'arts' subjects	36	20	—
Science	15	57	—
Other, or general studies	43	38	—
Weighted base	*170*	*90*	*20*

The columns of this table add to more than 100% since some visitors named more than one subject area.

Source of table: weighted results from count-based samples of museum leavers.

Table 2.6 Age at which full-time education was completed: visitors who had finished their full-time education

Finished education at age:	Victoria and Albert Museum	Science Museum	National Railway Museum
	%	%	%
16 or less	23	35	63
17–20	26	25	21
21 or over	51	39	17
Weighted base	*660*	*460*	*690*

Source of table: weighted results from count-based samples of museum leavers.

Table 2.7 Main subjects studied during last two years of full-time education by visitors who completed their education at the age of 17 or over

	Victoria and Albert Museum	Science Museum	National Railway Museum
	%	%	%
Visual arts	26	4	6
Other 'arts' subjects	35	29	38
Science	18	45	40
Other, or general studies	45	48	55
Weighted base	*500*	*300*	*260*

The columns in this table add to more than 100% since some visitors named more than one subject area.

Source of table: weighted results from count-based samples of museum leavers.

Table 2.8 Age at which full-time education was completed by museum visitors and by the general population

Finished education at age:	General population	Victoria and Albert Museum		Science Museum		National Railway Museum	
		Actual visitors	Expected visitors	Actual visitors	Expected visitors	Actual visitors	Expected visitors
	%	%	%	%	%	%	%
16 or less	79	30	75	40	75	66	76
17–20	13	25	15	26	15	19	13
21 or over	8	45	10	34	10	15	11
*Weighted base**	*—*	*420*	*—*	*350*	*—*	*620*	*—*

*Table refers to people aged 16–69, resident in Great Britain, who had finished their full-time education.

The figures for the general population are derived from the 1980 General Household Survey. The 'expected' figures for each museum are the figures that would have been obtained at that museum if visitors belonging to each of 14 categories (defined by sex and the age ranges 16–19, 20–24, 25–29, 30–39, 40–49, 50–59, and 60–69) had the same distribution of ages finished education as members of that category in the 1980 General Household Survey. The numbers belonging to each of the 14 categories, and the figures in the 'actual' columns, come from the weighted count-based samples of museum leavers.

The results shown in Table 2.5 indicate a marked though not unexpected contrast between the Victoria and Albert Museum—42 per cent of whose student visitors have specialised to a certain extent in some kind of visual art and only 15 per cent in a scientific subject—and the Science Museum, only eight per cent of whose student visitors have specialised in visual arts compared with 57 per cent who have specialised in science. In Chapter 3 we will investigate the extent to which this link between subjects studied and museum visiting comes about as a result of visits that are directly linked with courses of study and how far it results from a less specific interest in the student's general area of study.

The base line of Table 2.5 is equally interesting because of the information it gives about the proportion of each museum's visiting public who are students. The proportion of visitors who are full-time students aged 17 or above ranges from 17 per cent at the Victoria and Albert Museum down to two per cent at the National Railway Museum. The Science Museum occupies an intermediate position with nine per cent of its visitors being students.

Table 2.6 shows the ages at which visitors whose full-time education was complete had left school or college. Again the Victoria and Albert Museum seems to appeal particularly to a highly educated audience, 51 per cent of its visitors having continued their education up to or beyond their twenty first birthday. The visitors at both South Kensington museums are considerably more highly educated than those at the National Railway Museum, 63 per cent of whom finished their education before their seventeenth birthday.

Visitors who had continued their education past their seventeenth birthday were asked what were the main subjects they studied during their final two years of full-time education. Not surprisingly the results, shown in Table 2.7, correspond closely with the subjects studied by current students. Far more visitors to the Victoria and Albert Museum than to the other two museums had specialised in some form of visual arts though the proportion, at 26 per cent, was rather lower than the corresponding figure of 42 per cent for full-time students given in Table 2.5. The pattern of subject specialisation at the National Railway Museum resembles that at the Science Museum.

The proportion of highly educated people amongst the visitors of the two South Kensington museums seems from the evidence of Table 2.6 to be very much higher than that among the British population generally. In order to investigate this further Table 2.8 compares the results from this survey with findings from the General Household Survey, a major continuous survey carried out by the Office of Population Censuses and Surveys, whose results are representative of Great Britain as a whole. For the sake of comparability visitors aged 70 or over and visitors living in Northern Ireland or abroad have been excluded from the table.

The first column of the table gives figures for the population as a whole. The first column for each museum gives the distribution of ages at which its visitors finished their full-time education. The second column for each museum gives what that distribution would be if the museum's visitors were typical in this respect of other British people of the same sex and age. These figures differ slightly from the distribution for the population as a whole because school leaving ages have changed over time and because the age distribution of museum visitors is different from that of the population as a whole.

The results show that visitors to the two South Kensington museums are very much more highly educated than the British population generally. Forty-five per cent of the Victoria and Albert Museum's British visitors and 34 per cent of the Science Museum's British visitors continued their education to the age of 21 or beyond compared with only ten per cent which would be expected if the visitors to each museum were educationally typical of the whole population. The proportion of their visitors whose education ended before they were 17 is only about half what would be expected if they were typical of the public in general. The National Railway Museum's visitors, on the other hand, are only a little more highly educated than the population generally.

Table 2.9 shows another way in which the visitors of each museum are untypical of the country as a whole. All three museums are visited disproportionately by people living near them. Thus only 12 per cent of the population of the United Kingdom lives in Greater London, but 37 per cent of the Science Museum's and 59 per cent of the Victoria and Albert Museum's UK

Table 2.9 Regional distribution of UK resident visitors to each museum, compared to regional distribution of UK population

Resident in:	Victoria and Albert Museum	Science Museum	National Railway Museum	UK population 1981
	%	%	%	%
Greater London	59	37	3	12
Rest of South East	20	30	7	18
Yorks and Humberside	2	5	30	9
North and North West	4	6	25	17
Midlands and East Anglia	8	12	23	19
Wales and South West	5	10	4	13
Scotland	2	2	7	9
Northern Ireland	0	0	0	3
Weighted base	700	810	910	

Source of table: weighted results from count-based samples of museum leavers. The UK population figures are taken from Table 1 of Census 1981 Usual Residence, Great Britain *(CEN 81 UR, OPCS and Registrar General, Scotland) and from Table 3 of* The Northern Ireland Census 1981 Summary Report *and refer to people usually resident in the regions concerned.*

(The Census figures treat students differently from the survey's figures. The Census figures classify students to their home address. However museum survey respondents were simply asked for the name of the town and county they lived in; students were therefore free to give their term-time address if they wished.)

resident visitors live in the metropolis. Only nine per cent of the population, but 30 per cent of the National Railway Museum's UK resident visitors, live in the Yorkshire and Humberside region. Only three per cent of the National Railway Museum's visitors live in Greater London. (See also Figure 2.2.)

Table 2.10 shows the proportions of each museum's visitors who are residents of other countries. The figure is only nine per cent for the National Railway Museum but rises to 30 per cent in the case of the Victoria and Albert Museum. Later on in the report we will examine in detail the composition of this group.

Table 2.10   Percentages of visitors resident in the UK and abroad

	Victoria and Albert Museum	Science Museum	National Railway Museum
	%	%	%
Resident in UK	70	81	91
Resident abroad	30	19	9
Weighted base	1,000	1,000	1,000

Source of table: weighted results from count-based samples of museum leavers.

Most visitors at all three museums had some other recent experience of museum visiting. Well over half the visitors to each had been to two or more other museums or art galleries during the previous 12 months, as can be seen from Table 2.11. The Victoria and Albert Museum seems to have the highest proportion of museum buffs amongst its visitors, 30 per cent of whom, compared with only 13 per cent at the Science Museum and 11 per cent at the National Railway Museum, had visited ten or more other museums or galleries in the year before their interview. The proportion of visitors who had been to no other museum in the past 12 months was twice as high at the two technical museums as at the Victoria and Albert Museum.

Table 2.12 deals with previous visits to the museum in which the interview took place. Slightly less than half the visitors to each South Kensington museum were making their first visit ever, compared with 64 per cent who were visiting the newer National Railway Museum for the first time. Very few people had been to the National Railway Museum more than four times. At the other two museums however a considerable number of visitors could lay claim to something of a history of past visiting. Twenty-three per cent of the Science Museum's visitors had been there five or more times before, while the Victoria and Albert Museum again produces the highest proportion of real museum enthusiasts with 23 per cent of its visitors having made ten or more previous visits.

Table 2.11   The number of other museums and art galleries visited during the previous 12 months

Number of other museums and art galleries visited	Victoria and Albert Museum	Science Museum	National Railway Museum
	%	%	%
None	9	16	21
1	7	18	13
2–4	29	34	38
5–9	25	20	16
10 or more	30	13	11
Weighted base	1,000	1,000	1,000

Source of table: weighted results from count-based samples of museum leavers.

Table 2.12   Number of times visitors to each museum had visited it before

Number of previous visits	Victoria and Albert Museum	Science Museum	National Railway Museum
	%	%	%
None	47	46	64
1	8	12	12
2–4	13	19	18
5–9	9	10	3
10 or more	23	13	2
Weighted base	1,000	1,000	1,000

Source of table: weighted results from count-based samples of museum leavers.

c. Visiting alone and visiting with others

The previous section of this chapter showed marked differences between the segments of the population attracted by each of the three museums included in this study. These differences suggest the existence of corresponding differences in the type of experiences which people visiting each museum expect from their visits. Of course crude attendance figures by themselves provide a distorted image of each museum's appeal since they include both people who are visiting the museum for their own interest or enjoyment and people whose visit is not really due to any appeal the museum has for them personally. In this latter category will be found people who have come to the museum primarily to accompany someone else, for instance a friend or another member of their family, as well as many children visiting in school parties. At the other extreme solitary visitors are likely to be visiting because of their own interest. Data on the types of visitor who come to each museum alone or in various kinds of formal or informal group should therefore enable us to sharpen our picture of the particular public to which each museum really appeals.

Another reason for interest in the type of company in which people visit the museums is that it may well affect viewing patterns in ways that have implications for the design of the museums. This topic will be investigated later in the report, but the relevance for viewing patterns of the type of visiting group lends interest to the differing mixes of visiting groups at the three museums.

Table 2.13 gives a breakdown of the proportions visiting each museum alone and with different kinds of visiting group. A major distinction relating to who, if anyone, each visitor was accompanied by was whether or not the visitor had arrived at the museum in an organised party, for instance a party organised by a school, a college or a club of some kind. Family groups or groups of friends were not counted as organised parties—unless of course they happened to be visiting as part of a larger organised group. The proportions

of visitors arriving at each museum in school and other organised parties are set out in the fourth and fifth rows of Table 2.13. All told, visitors in organised parties account for 20 per cent or more of the Science and National Railway Museums' visiting public but for only 12 per cent of visitors to the Victoria and Albert Museum. The difference is due to the varying incidence of school party visiting which accounts for only six per cent of the Victoria and Albert Museum's public but for 19 per cent of visitors to the Science Museum and 14 per cent at the National Railway Museum.

Tables 2.14 to 2.16 show, for each museum separately, the proportions of different age groups visiting with organised parties. Table 2.15 shows that school parties account for about 30 per cent of the Science Museum's children and teenaged visitors. At the Victoria and Albert Museum the figure is 17 or 18 per cent but at the National Railway Museum it rises to nearly 40 per

Figure 2.2 Regional distribution at each museum (from Table 2.9)

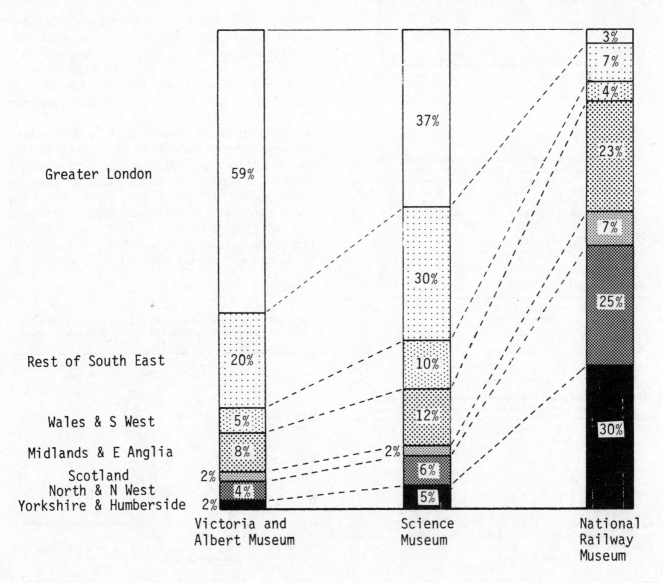

Table 2.13 Proportions of visitors coming to each museum alone and with different kinds of visiting group

Visiting:	Victoria and Albert Museum	Science Museum	National Railway Museum
	%	%	%
Alone	26	10	5
With friends but not members of family	24	20	11
With members of family	38	45	64
With a school party	6	19	14
With another type of organised party	6	6	6
Weighted base	*1,000*	*1,000*	*1,000*

Source of table: weighted results from count-based samples of museum leavers.

Table 2.14 Proportions of different age groups visiting with organised parties at the Victoria and Albert Museum

Age (years)		With school party	With other organised party	Not with organised party	*Weighted base*
Up to 10	%	18	0	82	*60*
11–20	%	17	10	72	*230*
21–30	%	1	5	95	*280*
31–40	%	2	6	93	*160*
41 or over	%	0	4	95	*270*
All ages	%	5	6	89	*1,000*

Source of table: weighted results from count-based sample of museum leavers.

Table 2.15 Proportions of different age groups visiting with organised parties at the Science Museum

Age (years)		With school party	With other organised party	Not with organised party	*Weighted base*
Up to 10	%	29	4	67	*190*
11–20	%	33	10	57	*360*
21–30	%	5	2	93	*150*
31–40	%	3	3	94	*150*
41 or over	%	1	4	95	*140*
All ages	%	19	6	75	*1,000*

Source of table: weighted results from count-based sample of museum leavers.

Table 2.16 Proportions of different age groups visiting with organised parties at the National Railway Museum

Age (years)		With school party	With other organised party	Not with organised party	*Weighted base*
Up to 10	%	41	10	50	*110*
11–20	%	37	8	54	*210*
21–30	%	3	4	93	*150*
31–40	%	3	5	92	*270*
41 or over	%	1	6	93	*260*
All ages	%	14	6	80	*1,000*

Source of table: weighted results from count-based sample of museum leavers.

cent. Indeed almost half the under twenty ones coming to the National Railway Museum visit it as part of some kind of organised group. Teachers and other adults accompanying school parties account for a few of the older visitors to each museum.

College parties and visits by youth organisations such as wolf cubs mean that visits in the 'other organised parties' category are also relatively common in the under 21 age group. (Visits by parties of foreign school children were also included in the 'other organised parties' category.)

The remaining tables in this section exclude organised parties in order to concentrate on the visitors who come to each museum alone or in informal groups of family members or friends. Table 2.13 shows what proportions these represent of the whole visiting public. Other informal groups, for instance groups of colleagues, were very rare. No one belonging to a group of this sort was interviewed at either the Victoria and Albert Museum or the National Railway Museum, while less than half of one per cent of the weighted sample at the Science Museum were visiting with colleagues. In Table 2.13 and subsequent tables they have been grouped with people visiting in the company of friends.

Table 2.17 shows that the National Railway Museum is overwhelmingly a family museum. Eighty per cent of its non-organised visitors came with members of their families. The same is true to a lesser degree of the Science Museum. Family visits play a smaller role at the Victoria and Albert Museum, though even there 43 per cent of the non-organised visitors had come with members of their families. (In fact there is also a difference between the type of family group at each museum. At the Science Museum 73 per cent of people visiting with members of their families had come as part of a parent-child group, that is a group containing one or more people related to the interviewee as parent or child. Thirteen per cent of the Science Museum's family visitors were visiting with their spouse only, and 13 per cent were visiting with some other grouping of

Table 2.17 Proportions of non-party visitors coming to each museum alone, with friends or with members of their families

	Victoria and Albert Museum	Science Museum	National Railway Museum
	%	%	%
Alone	30	13	6
With friends but not members of family	27	26	14
With members of family	43	60	80
*Weighted base**	*880*	*750*	*800*

Excludes members of organised parties.

Source of table: weighted results from count-based samples of museum leavers.

relatives. The proportion of family visitors accompanied by their spouse only was slightly higher, 19 per cent, at the National Railway Museum but reached 33 per cent at the Victoria and Albert Museum.)

When we turn to look at the proportions of lone visitors shown in Table 2.17, we see that the Victoria and Albert Museum has the highest proportion with 30 per cent followed by the Science Museum with 13 per cent and the National Railway Museum with only six per cent. If it is right that the kind of visiting group accompanying an individual to the museum influences the whole pattern of his visit, the different mixes of visiting groups at the three museums suggest, quite as strongly as differences in the social characteristics of the visitors themselves, that people come to the three museums in search of qualitively different experiences. It seems likely that whatever it is about the Victoria and Albert Museum that attracts especially a rather intellectual and museum-oriented public goes best with a quiet solitary style of visiting. Far fewer visitors to the Science and National Railway Museums appear, on the face of it, to fear that company will detract from the kind of experience they have come for.

The final four tables of this chapter take each museum separately and look in more detail at who visit with their families, who come to each museum with their friends and who visit the museums on their own. Tables 2.18, 2.19 and 2.20 relate these different visiting groups to age. The pattern that emerges is very similar for all three museums reflecting, as it does, the changes that occur in individuals' social lives as they pass through different age bands. The overwhelming majority of children under 11 are accompanied by members of their families. This proportion falls to a much lower level among the 11–20s. As people reach an age to form their own families a reverse movement occurs. There is a dramatic jump in the proportion of family visitors as we move from the 21–30 to the 31–40 age group by which age many visitors, particularly at the Science and National Railway Museums, are bringing children of their own. (Of the family visitors aged 31–40, 88 per cent at the Science Museum and 90 per cent at the National Railway Museum were in fact part of a parent-child group.) The 11–20 and 21–30 bands are the great ages for visiting with friends. Visiting alone is very rare in the under 11 age band but otherwise appears to have little relation to age.

As we noted in the previous section of this chapter the 11–20 age group is a very varied one. Table 2.21 therefore focuses in detail on the different segments that make it up. The figures in the table show that there is a preponderance of family visiting in the case of school children aged 16 or less at all three museums. Those visiting without members of their family are almost always accompanied by friends. From the mid-teens the pattern changes. The proportion visiting with friends rises to 50 per cent or more, the number of solitary visitors increases greatly while, among student visitors at the two South Kensington museums, the proportion visiting with their families falls to almost nothing.

Table 2.18 Proportions of non-party visitors coming to the Victoria and Albert Museum alone, with friends or with members of their families by sex and age

Sex and age (years)		Alone	With friends but not members of family	With members of family	Weighted base*
Males	%	33	22	45	370
Females	%	27	31	42	510
Persons					
Up to 10	%	0	0	100	50
11–20	%	21	47	32	170
21–30	%	37	34	29	260
31–40	%	35	17	49	150
41 or over	%	30	19	51	260
All non-party visitors	%	30	27	43	880

Excludes members of organised parties.
Source of table: weighted results from the count-based sample of museum leavers.

Table 2.19 Proportions of non-party visitors coming to the Science Museum alone, with friends or with members of their families by sex and age

Sex and age (years)		Alone	With friends but not members of family	With members of family	Weighted base*
Males	%	18	29	53	490
Females	%	5	23	72	260
Persons					
Up to 10	%	2	9	89	130
11–20	%	13	49	39	210
21–30	%	16	39	45	140
31–40	%	12	14	74	140
41 or over	%	25	8	68	140
All non-party visitors	%	13	26	60	750

Excludes members of organised parties.
Source of table: weighted results from the count-based sample of museum leavers.

Table 2.20 Proportions of non-party visitors coming to the National Railway Museum alone, with friends or with members of their families by sex and age

Sex and age (years)		Alone	With friends but not members of family	With members of family	Weighted base*
Males	%	9	19	72	480
Females	%	1	7	92	310
Persons					
Up to 10	%	0	0	100	60
11–20	%	9	36	55	120
21–30	%	3	31	66	140
31–40	%	5	4	91	250
41 or over	%	8	8	84	240
All non-party visitors	%	6	14	80	800

Excludes members of organised parties.
Source of table: weighted results from the count-based sample of museum leavers.

At the Victoria and Albert Museum Table 2.18 shows that the proportions of solitary and accompanied visitors are much the same for each sex. However Tables 2.19 and 2.20 show that far fewer women than men visit the Science Museum and the National Railway Museum alone, while far more of the two technical museums' women visitors are accompanied by other members of their families. This indicates that these two museums' relatively greater appeal to men and boys, which was apparent from the attendance figures in Section (b) of this chapter, is even more marked than those figures suggest. It looks as though many of the women who do visit the Science and National Railway Museums are simply following in the wake of husbands or sons. (These results do not necessarily imply that women find the two museums less enjoyable than men do once they are there—see Chapter 12.)

Table 2.21 Proportions of 11–20 year old visitors coming to each museum alone, with friends or with members of their families by status—non-party visitors

		Alone	With friends but not members of family	With members of family	Weighted base*
Victoria and Albert Museum					
School children aged 11–16	%	5	28	66	70
Aged 17–20 in full-time education	%	33	63	4	70
Finished full-time education	%	30	49	21	30
Science Museum					
School children aged 11–16	%	7	37	57	120
Aged 17–20 in full-time education	%	26	69	5	50
Finished full-time education	%	17	64	19	30
National Railway Museum					
School children aged 11–16	%	5	30	65	80
Aged 17–20 and/or finished full-time education	%	20	52	28	30

*Excludes members of organised parties.

Source of table: weighted results from the count-based samples of museum leavers.

3 Why people visit the museums

a. General

In this chapter we will make use of data derived from both the count-based and the quota-based samples of people leaving the museums. It is worth recapping briefly what was said in Chapter 1 about the strengths and weaknesses of the two types of sample. The count-based samples are (except for a few qualifications discussed in Appendix 1) fully representatives of each museum's visitors but, because of the need to keep the questionnaire brief, visitors to the two London museums could be asked only one question each about their reasons for visiting the museum. Visitors to the National Railway Museum were asked, in addition, why they were visiting at the particular time of year. We were able to collect much fuller information from the visitors who agreed to take part in the longer interviews that were carried out on the days when quota sampling was used. However we have to bear in mind some qualifications when using the quota sampled data. There is the possibility, discussed in Appendix 1, that the visitors interviewed because they belong to a specific category are not fully representative of all visitors in that category. There is also the point that, in order to have adequate numbers for analysis in various categories, the quotas were deliberately designed to 'over-sample' certain types of visitor and 'under-sample' others. Finally it is important to remember that, for reasons of convenience, we excluded certain types of visitors altogether from the quotas of people leaving the museums—specifically under-elevens, visitors with organised parties, and people who did not look at any of the exhibits because they were visiting the museum purely to use its library, attend a lecture or for some other reason. The differing composition of the three museums' visiting publics, dicussed in the previous chapter, means that the decision to exclude under-elevens and people visiting in organised parties from the quotas will have ruled out fewer visitors at the Victoria and Albert Museum than at the two technical museums.

These qualifications do not render the data selected from the quota samples invalid. However the quota-based results do need to be treated with a little more caution than those based on the count-linked samples. It is particularly important at points in the discussion where data from both kinds of sample are used, to keep in mind the different ways in which the information was collected.

b. Deciding to visit

Interviewers asked visitors in the count-based sample what had been their main reason for deciding to visit the museum. Table 3.1 reports their answers or, in the

Table 3.1 The reasons given for visiting each museum

Reasons for visiting	Victoria and Albert Museum	Science Museum	National Railway Museum
	%	%	%
To accompany someone else	11	33	40
General interest, reputation of the museum	21	18	19
Because the visitor liked the museum	5	8	7
Interest in the museum's contents	25	27	37
To see a special temporary exhibition	26	10	0
In connection with work or studies	7	5	3
To use information facilities	5	1	0
Casual or holiday visit	9	13	21
Other	12	8	4
Weighted base	*1,000*	*1,000*	*1,000*

The percentages add to over 100% as people could give more than one answer.

Source of table: weighted results from count-based samples of museum leavers.

case of small children and visitors unable to speak English, the answers made on their behalf by people with them. A number of visitors offered more than one 'main reason'; all the reasons they volunteered have been included in the table.

Many of the visitors at the Science and National Railway Museums were there to accompany someone else (Table 3.2). They included people who were visiting the museum because friends or other members of their families wanted to see it and people who said they had come to the museum because they belonged to a group who were there in an organised party. The most striking difference between the reasons given by visitors to the Victoria and Albert Museum and those offered by visitors to the two technical museums is that nearly all the Victoria and Albert Museum's visitors seem to have been there mainly for their own interest.

Part of the explanation of this difference is the fact, noted in the last chapter, that children in school parties make up a considerable proportion of visitors at the Science and National Railway Museums. Tables 3.3 and 3.4 show that nearly half the teachers and pupils visiting these two museums in school parties seem to have felt that the main reason they were there was simply the fact that the party had been organised.

Far more important, however, are the Science and National Railway Museums' roles as 'family museums'. The need to accompany someone else was cited as a main reason by over 40 per cent of the people visiting these two museums with other members of their families. People visiting with friends were much more likely to have wanted anyway to come to the museum themselves.

One result of this pattern of 'altruistic' family visiting is that the likelihood of a visitor to the Science or National Railway Museum being there primarily because he or she really wants to visit it varies consider-ably according to the visitor's age. Table 3.5, which excludes organised parties, shows that half these museums' 31–40 year old visitors—the prime parental age group—had gone there to take someone else. Altruistic visiting is much less common in the 11–20 age group and, at the Science Museum in the 21–30 age group also. It is rather more common with visitors aged ten or less, though in their case it might perhaps be better described as 'compulsory' visiting.

Table 3.2 Reasons given for visiting the Victoria and Albert Museum by people visiting it alone or in a group

Reasons for visiting	Visiting:				
	Alone	With friends but not members of family	With members of family	With a school party	With another type of organised party
	%	%	%	%	%
To accompany someone else	1	7	15	36	23
General interest, reputation of the museum	15	22	26	15	8
Because the visitor liked the museum	5	5	5	1	2
Interest in the museum's contents	23	19	26	22	46
To see a special temporary exhibition	28	40	21	9	11
In connection with work or studies	9	3	2	12	35
To use information facilities	12	3	1	6	3
Casual or holiday visit	10	6	11	6	5
Other	17	10	11	8	7
Weighted base	*260*	*240*	*380*	*60*	*60*

The percentages add to over 100% as people could give more than one answer.
Source of table: weighted results from count-based samples of museum leavers.

Table 3.3 Reasons given for visiting the Science Museum by people visiting it alone or in a group

Reasons for visiting	Visiting:				
	Alone	With friends but not members of family	With members of family	With a school party	With another type of organised party
	%	%	%	%	%
To accompany someone else	0	12	44	43	51
General interest, reputation of the museum	11	23	20	14	22
Because the visitor liked the museum	15	9	9	1	2
Interest in the museum's contents	35	27	23	30	34
To see a special temporary exhibition	16	14	9	5	7
In connection with work or studies	8	2	1	16	10
To use information facilities	3	1	0	2	0
Casual or holiday visit	19	20	12	7	0
Other	16	9	7	5	0
Weighted base	*100*	*200*	*450*	*190*	*60*

The percentages add to over 100% as people could give more than one answer.
Source of table: weighted results from count-based samples of museum leavers.

Table 3.4 Reasons given for visiting the National Railway Museum by people visiting it alone or in a group

Reasons for visiting	Visiting:				
	Alone	With friends but not members of family	With members of family	With a school party	With another type of oranised party
	%	%	%	%	%
To accompany someone else	0	20	43	50	50
General interest, reputation of the museum	16	23	22	4	13
Because the visitor liked the museum	25	5	8	3	0
Interest in the museum's contents	43	51	36	33	25
To see a special temporary exhibition	5	1	0	0	0
In connection with work or studies	4	2	0	16	7
To use information facilities	0	0	0	2	0
Casual or holiday visit	23	25	21	15	20
Other	16	4	4	4	3
Weighted base	*50*	*110*	*640*	*140*	*60*

The percentage add to over 100% as people could give more than one answer.
Source of table: weighted results from count-based samples of museum leavers.

The proportion of under-elevens visiting because others were visiting is higher at the Science Museum than at the National Railway Museum. But it is highest of all in the Victoria and Albert Museum, reflecting the lack of appeal to small children which makes the museum less of a focus for family visits than are the other two. Those who do visit the Victoria and Albert Museum with other members of their family are generally interested on their own account. Table 3.2 records that only 15 per cent of the Victoria and Albert Museum's family visitors said they were there to accompany the others.

Table 3.5 The percentages of non-party visitors in each age range who visited each museum in order to accompany someone else

Age (years)	Victoria and Albert Museum (a)	Science Museum (b)	National Railway Museum (c)	Weighted bases		
				(a)	(b)	(c)
Up to 10	49%	39%	27%	50	130	60
11–20	6%	13%	19%	160	210	120
21–30	4%	19%	37%	260	140	140
31–40	10%	48%	52%	150	140	250
41 or over	7%	39%	32%	260	140	240
All ages	9%	30%	37%	880	750	800

Source of table: weighted results from count-based samples of museum leavers.

Returning to Table 3.1 we see that about 20 per cent of each museums visitors gave general interest or the reputation of the museum as a reason for being there. Typical answers were: *"Just to look round"*, *"Just to see the place, I haven't been before"*, *"It's a tourist attraction"* and *"I was told by a friend it was well worth a visit"*.

Rather more visitors, particularly in the National Railway Museum, said that they were interested in the museum's contents—referring either to the general subject area or to specific galleries or exhibits. About a quarter of the Victoria and Albert Museum's visitors cited a special temporary exhibition as a 'main reason' for their visit, as did ten per cent at the Science Museum but less than half a per cent at the National Railway Museum.

More visitors at the National Railway Museum than at the two London museums said they were there on a casual or holiday visit. Answers placed in this category included: *"It was somewhere to go"*, *"We are in York"*, *"We are on holiday, it's our first time in London"*, *"Just a day out"* and *"We were in the area"*.

The proportion of visitors giving their work or studies as a main reason for their visit ranged from three per cent at the National Railway Museum to seven per cent at the Victoria and Albert Museum. Five per cent of its visitors said they were visiting the Victoria and Albert Museum to use its information facilities—by attending a lecture, using the library or bringing in an

object for an opinion from the museum's staff—but this was seldom volunteered as a reason for visiting either of the technical museums.

Tables 3.2 to 3.4 show that solitary visitors to the different museums are more likely than other non-party visitors to be there for reasons connected with their work or studies. However the link with studies is most often cited by visitors in school parties and other organised parties.

(Note that teachers and guides who said they were visiting the museum in order to accompany organised parties were not categorised as 'visiting in connection with their work' but rather as 'accompanying others'.)

The pattern of replies recorded in Tables 3.2 to 3.4 illustrates an important point about the answers visitors gave to our question about their 'main reason' for visiting the museum. This is that the 'main reason' need not be the only important reason. The proportion of school party visitors giving the existence of the school party as a 'main reason' for their visit does not rise above 50 per cent despite the fact that, but for the party visit, hardly any of its members would be at the museum at that time. It may also be that not everyone visiting in connection with their work or studies, or visiting the museum because of an interest in a particular subject, mentioned the fact at this question. Because of this problem the rest of this chapter will place less emphasis on the question about people's 'main reason' for visiting. Instead the discussion will largely be based on answers to the more specific questions put to the quota samples of leavers.

c. Differing priorities
In the previous chapter we saw that a substantial majority of visitors to the two technical museums were male, and that when women or girls did visit either museum they were very likely to be with other members of their families. Table 3.6 completes the picture by showing that they were considerably more likely than men and boys to give 'accompanying someone else' as their main reason for visiting the two museums. They were much less likely then men and boys to cite an interest in either museum's contents. There is virtually no difference between the reasons given by women for visiting the Victoria and Albert Museum and those given by men.

It was suggested in the previous chapter that the different composition of the museums' visiting publics in terms of age, level of education and the proportion of solitary visitors might reflect differences in the kind of experiences visitors expected from their visits. The underlying assumption was that these different sorts of visitors would be looking for different things. Tables 3.7 to 3.9 confirm that these characteristics are associated with differences in people's attitudes to museum visiting.

The tables are based on the following question, put to the quota samples of leavers:

Table 3.6 The reasons given for visiting each museum by sex

Reasons for visiting	Victoria and Albert Museum		Science Museum		National Railway Museum	
	Males	Females	Males	Females	Males	Females
	%	%	%	%	%	%
To accompany someone else	12	10	29	41	30	55
General interest, reputation of the museum	21	21	17	21	20	17
Because the visitor liked the museum	4	4	8	7	8	5
Interest in the museum's contents	25	25	32	17	44	26
To see a special temporary exhibition	24	28	11	9	0	0
In connection with work or studies	7	6	4	7	3	3
To use information facilities	4	5	1	1	0	1
Casual or holiday visit	10	7	14	10	21	20
Other	12	12	7	9	4	4
Weighted base	430	570	630	360	610	390

The percentages add to over 100% as people could give more than one answer.

Source of table: weighted results from count-based samples of museum leavers.

"When you have been round a museum is it important to you to feel that you have learnt something or do you just enjoy looking at things?"

In all three museums older visitors are notably keener on learning than are younger visitors. There is a similar, but less pronounced, difference between people whose full-time education continued to the age of 21 or beyond and those who completed it before their seventeenth birthday. The picture revealed by Table 3.9, however, is rather more complex. In all three museums solitary visitors are more likely than people visiting with friends or family to feel it is important to learn something. However it is surprising to find that people visiting in family groups are keener on learning than people visiting with friends. The explanation lies in the fact that the quotas excluded children under eleven. Families tend, therefore, to be represented in these tables by parents or people visiting just with their spouses. These people share the general interest that older visitors show in the idea of learning something—even if a family visit is not the context in which they themselves would learn most.

Table 3.7 Percentages of visitors who thought learning was important by age

Age (years)	Victoria and Albert Museum (a)	Science Museum (b)	National Railway Museum (c)	Bases		
				(a)	(b)	(c)
11–20	43%	47%	36%	255	337	187
21–30	54%	57%	41%	262	320	198
31–40	60%	64%	44%	131	167	82
41 and over	62%	69%	60%	355	221	129

Source of table: the quota samples of museum leavers.

Table 3.8 Percentages of visitors who thought learning was important by age at which full-time education was completed

Age at which education was completed	Victoria and Albert Museum (a)	Science Museum (b)	National Railway Museum (c)	Bases*		
				(a)	(b)	(c)
16 or less	50%	53%	44%	172	271	258
17–20	55%	57%	48%	246	313	135
21 or over	63%	65%	54%	318	245	77

*Visitors still in full-time education are excluded from this table.

Source of table: the quota samples of museum leavers.

Table 3.9 Percentages of visitors who thought learning was important by whether they came to the museums alone, with friends or with members of their families

	Victoria and Albert Museum (a)	Science Museum (b)	National Railway Museum (c)	Bases		
				(a)	(b)	(c)
Alone	58%	67%	52%	361	287	126
With friends but not members of family	51%	45%	39%	280	320	155
With members of family	55%	61%	43%	362	438	315

Source of table: the quota samples of museum leavers.

d. What visitors came to see
i. General

Members of the quota samples of people interviewed leaving each museum were asked whether they, or people with whom they had come to the museum, had set out for the museum with the aim of seeing anything particular. Fewer visitors at the National Railway Museum than at the other two museums had anything specific in mind to see, as the figures in Table 3.10 show. Of course the meaning of the question was somewhat different at the National Railway Museum, where it went without saying that people had come to see exhibits concerned with railways, than at the two London Museums where many visitors answered in terms of which subject areas they had come to see.

Table 3.10 Whether visitors (or their companions) had decided before arriving at the museum that there was something they definitely wanted to see

	Victoria and Albert Museum	Science Museum	National Railway Museum
	%	%	%
Came to see something definite	62	48	29
Did not come to see anything definite	38	52	71
Base	1,003	1,045	596

Source of table: the quota samples of museum leavers.

ii. The Victoria and Albert Museum

When visitors to the Victoria and Albert Museum were asked what it was that they had definitely wanted to see they divided into the two roughly equal groups shown in Table 3.11. About half referred to a special temporary exhibition or, occasionally, named a specific part of the museum's permanent collection such as the Jones collection without specifying the type of exhibit that interested them. However nearly everyone who had come to see something specific other than a temporary exhibition gave the second kind of answer—that is they explained what sort of exhibits they had come to see. Since a few visitors had arrived with the definite intention both of seeing a specific exhibition and of looking in a more general way at some aspects of the museum's collection the figures in Table 3.11 add to over 100 per cent.

Table 3.11 How visitors to the Victoria and Albert Museum described what they or their companions wanted to see

	%
Named a temporary special exhibition or a specific permanent gallery	49
Described the things they had come to see	55
Base	625

The percentages add to over 100% as some people gave answers of both types.

Source of table: the quota sample of museum leavers.

Table 3.12 Exhibitions which visitors to the Victoria and Albert Museum had come to see

	%
Permanent gallery	11
Hollywood	16
Japan Style	30
Princely Magnificence	16
McCullin photographs	12
Other temporary special exhibitions	22
*Base**	304

*All visitors who named a temporary exhibition or a specific permanent gallery.

The percentages add to over 100% as people could give more than one answer.

Source of table: the quota sample of museum leavers.

Table 3.12 lists, for future reference, the exhibitions which had attracted considerable numbers of the visitors included in our quota sample. The exhibitions' role in attracting different kinds of visitor to the museum will be discussed in Chapter 8. The figures in the table do not indicate the relative popularity of the exhibitions since the survey period coincided with different proportions of each exhibition's life.

It is Table 3.13 which tells us which aspects of the museum's permanent collections visitors particularly came to see. The table is divided into four parts dealing with respectively the particular art forms, their geographical origins, the period at which they were made and the artist who made them. Thus the answer of a visitor who said that he had come to look at paintings by Constable would be coded as 'paintings' under 'art form' and 'Constable' under 'artist'. The answer would be coded 'not specified' in the sections on geographical origin and the period the work was done, since these were not mentioned specifically by the visitor. If the visitor had also decided to look at Chinese pottery, that answer too would be entered in Table 3.13 (as 'ceramics' under 'art form', 'Chinese' under 'geographical origin' and 'not specified' under 'period' and 'artist'). The coding of the answers into these four aspects was done clerically after the interviewing was completed. The interviewers themselves accepted the answers in the form that they were given by the interviewees.

Table 3.13 Types of exhibit visitors came to see at the Victoria and Albert Museum (excluding special exhibitions)

Art form	%	Geographical origin	%
'Art'	7	British	5
Paintings	17	European	4
Portrait miniatures	2	Islamic	3
Prints	1	Indian	3
Sculpture	5	Chinese, Japanese, Korean	6
Casts	0	Oriental (country not specified)	2
Furniture	13	South East Asian	0
Woodwork	1	Other	1
Ceramics	8	Geographical origin not specified	84
Tapestries	3	**Period**	**%**
Carpets	2	16th Century	1
Textiles	3	17th Century	2
Embroidery, needlework	6	18th Century	2
		Renaissance	2
Costumes	23	Victorian	3
Stained glass	1	Tudor	1
Musical instruments	8	Other	4
Glass vessels	2	Period not specified	91
Silver	2	**Artist**	**%**
Jewellery	7	Constable	13
Metalwork	0	Turner	1
Ironwork	1	William Morris	1
Armour, weaponry	1	Raphael	4
Other	8	Other	2
Art form not specified	13	Artist not specified	85

Base* = 342

*Visitors are only included in this table if they explained what kinds of exhibits they, or their companions, had decided to see.

The percentages in each section add to over 100% as people could give more than one answer.

Source of table: the quota samples of museum leavers.

A comparison of the four sections of the table indicates that visitors came to see particular art forms rather than work done in particular parts of the world, by particular artists or at particular historical periods. Only 13 per cent of the visitors who explained the kinds of exhibit they had come to see defined any type of exhibit in terms of its geographical origin, of who made it or of when it was made without also mentioning the art form involved. By contrast 84 per cent managed to describe at least one type of exhibit without mentioning any particular geographical origin, 85 per cent managed it without mentioning the artist and 91 per cent without mentioning the historical period. In other words it was common for people to say simply that they had come to look at pottery, ceramics, painting or some other art form but unusual for them to say simply that they had come to see Indian, Chinese, European or British work (or to mention the period or the artist) without specifying the art form involved. Visitors seem to have found it most natural to define their specific interests in terms of of particular forms of art or craft rather than on any other basis.

Costumes were the art form most often cited by the visitors—even though the costume gallery was closed for repairs throughout the survey year. Paintings came next and furniture after that. Musical instruments, ceramics, jewellery and embroidery were each mentioned by six to eight per cent of the visitors who explained the type of exhibit they had planned to see. Despite the museum's considerable collection of sculpture only five per cent mentioned it as something they had decided to see before their arrival at the museum.

Constable received far more mentions than any other individual artist—in fact 13 per cent of the people who had come to the museum to see a particular kind of exhibit were there specifically to see the Constables. Of the few visitors who had come specifically to see foreign work, most mentioned oriental things. It would probably be incorrect to conclude that this reflects a lack of interest in European art. It is more likely that European work is so much part of visitors' ideas of art that its geographical origin is not worth mentioning. Hardly anyone had come to the museum specifically to see Victoriana or indeed the work of any other particular period.

iii. The Science Museum and the National Railway Museum

The Science Museum is divided into galleries which correspond fairly clearly to divisions in subject matter. It was therefore possible to tell, on the basis of the things visitors said they had decided to see, which galleries were of particular interest to them. Table 3.14 can therefore be thought of as referring either to galleries or to different subjects.

Five subjects, or galleries, in the museum's permanent collection seem to have played a particularly notable part in people's plans to visit the museum. These are the Children's gallery whose importance would no doubt have been even clearer if children aged ten or less had been included in the quota samples; road and rail transport; computing; aircraft; and space and rocketry. Thirty-one per cent of the visitors who had planned in advance to look at something specific mentioned other galleries in the permanent collection but no other individual gallery was mentioned by more than three per cent. There seems to be a very clear difference between the drawing power of what might be called the 'star attractions' and that of the remaining galleries.

Two special temporary exhibitions had played a notable part in people's plans to visit the museum. These were 'The Challenge of the Chip'—an exhibition devoted to the many applications of micro-chip technology—which ran for almost the whole of the survey year, and 'The Great Optical Illusion' which was concerned with the development of television broadcasting and ran from the end of March to the end of September.

Table 3.15 lists the items visitors to the National Railway Museum particularly intended to see. The star

Table 3.14 What visitors came to see at the Science Museum

Permanent galleries	%
Road and rail transport	14
Aircraft	13
Children's gallery	9
Space and rockets	9
Computing	8
Other permanent galleries	31
Special exhibitions	
The Challenge of the Chip*	20
The Great Optical Illusion†	12
Other special temporary exhibitions	4
*Base*ø	505

*Open for all but the first six days of quota interviewing.

†Open for six months during the survey year.

øVisitors are only included in this table if they, or people accompanying them, had decided before arriving at the museum that there was something they definitely wanted to see.

Percentages add to over 100% as people could give more than one answer.

Source of table: the quota samples of museum leavers.

Table 3.15 The things visitors came to see at the National Railway Museum

	%
Mallard	37
*Rocket**	9
APT, High Speed Trains	10
Specific locomotive or class of locomotives	11
Locomotives but not precisely specified	20
Royal coaches	17
Coaches (not specifically royal)	6
Other	24
Base†	173

*The Rocket itself was not housed at the National Railway Museum. The museum did have a newly constructed full-size reproduction in working order, though it was not on display for some of the period covered by the quota interviews.

†Visitors are only included in this table if they, or people accompanying them, had decided before arriving at the museum that there was somthing they definitely wanted to see.

Percentages add to over 100% as people would give more than one answer.

Source of table: the quota samples of museum leavers.

attraction was the streamlined *Mallard* locomotive. A number of visitors had come to see Stephenson's rocket a full-scale working model of which was on show in the Main Hall during some at least of the survey period (though it arrived too late to be included in our detailed plan of the Main Hall). About the same number were particularly interested in the Advanced Passenger Train or high speed trains generally.

Several other visitors mentioned particular locomotives or classes of locomotive and 20 per cent of the visitors who had something they particularly wanted to see referred in general terms to diesels, steam locomotives or simply said that they had come to see the engines. Seventeen per cent had come to see the royal coaches and another six per cent mentioned coaches without specifically singling out the royal ones.

e. Links with visitors' work or studies

Table 3.16 shows, separately, the percentages of each museum's visitors who gave their job or their studies as a main reason for visiting the museum. (The fact that visitors might give both reasons and the convention followed in this report of rounding percentages to the nearest whole number account for the apparent slight discrepancy between these figures and those for the combined category in Table 3.1.) Table 3.17 gives the equivalent percentages for the quota samples of leavers—which of course excluded visitors who went to the museums purely to use the information services as well as visitors in school or other organised parties.

Table 3.18, again based on the quota samples, shows the larger, though still modest, percentages who when asked specifically agreed that the visit would help them with their job or with their studies. The special value of what these people have to say is that it should show how looking round the public exhibition galleries themselves (rather than, say, just visiting the museum's library) can help with work and study.

As Table 3.19 indicates most of those who felt that the visit would help with their jobs thought that it would do so by broadening their understanding. The answers placed in this category included both references to information on precise subject areas and wider references to the acquisition of general insight but did not include references to specific applications of the understanding gained. Visitors cited two main ways in which the knowledge acquired could actually be applied in their jobs. Most common were answers from teachers and others who said that their present visit gave them the background information they needed to conduct a subsequent visit by an organised party, or provided knowledge that could help in some other way with their teaching. Less common were replies from a number of people who, particularly at the Victoria and Albert Museum, saw the visit as a source of practical ideas. An example was the visitor who said *"I find specific motifs and then I develop them for my own use"*.

Table 3.16 Percentages of visitors who cited their job or studies as a main reason for visiting each museum

Main reasons included	Victoria and Albert Museum	Science Museum	National Railway Museum
Job	2%	0%	1%
Studies	4%	5%	3%
Weighted base	*1,000*	*1,000*	*1,000*

Source of table: weighted results from the count-based sample of museum leavers.

Table 3.17 Percentages of quota sampled visitors who cited their job or studies as a main reason for visiting the museum

Main reasons included	Victoria and Albert Museum	Science Museum	National Railway Museum
Job	1%	2%	1%
Studies	2%	3%	1%
Base	*1,003*	*1,045*	*596*

Source of table: the quota samples of museum leavers.

Table 3.18 Percentages of quota sampled visitors who thought the visit would help with their job or studies

Visit would help with:	Victoria and Albert Museum	Science Museum	National Railway Museum
Job	7%	6%	1%
Studies	8%	8%	2%
Base	*1,003*	*1,045*	*596*

Source of table: the quota samples of museum leavers.

Table 3.19 How the visit would help visitors with their jobs

Visit would:	Victoria and Albert Museum	Science Museum	National Railway Museum
	%	%	
Broaden understanding	59	56	—
Prepare way for bringing party of pupils/provide knowledge that could be passed on to others	27	39	—
Provide practical ideas	17	8	—
Help in other ways	4	5	—
Base	*70*	*64*	*6*

The percentages add to over 100% as people could give more than one answer.
Source of table: the quota samples of museum leavers.

In Chapter 2 we noted a marked contrast between the subjects studied by the student visitors of the two London Museums and wondered whether this was a direct consequence of study-linked visiting or whether the differing proportions of students of science on the one hand and visual arts on the other among visitors to the two museums resulted simply from the greater appeal of the Victoria and Albert Museum to the artistically minded and of the Science Museum to the scientifically inclined. The following tables, like those in the previous chapter, define a student as anyone aged 17 or over who is in full-time education. The following tables suggest that about half the visits to the Victoria and Albert Museum by students who gave some form of visual art as a main subject had some kind of educational link. The same appears to be true of visits to the Science Museum by students of science.

School parties and other organised parties were responsible for about a quarter of the visits by science students to the Science Museum and for the same proportion of visits by visual arts students to the Victora and Albert Museum, as the figures in Tables 3.20 and 3.21 show. It is likely that many of the 'other organised parties' were college parties or had some other connection with education.

If we turn to students visiting alone or in informal groups, we see from Tables 3.22 and 3.23 that 17 per cent of the visual art students at the Victoria and Albert Museum and 18 per cent of the science students at the Science Museum gave their studies as a main reason for the visit. However figures from the quota samples of leavers—which are of course themselves confined to people visiting on their own or with friends or family—suggest that the proportion of non-party student visitors who thought that their visit would help

Table 3.20 Percentages of full-time students aged 17 and over visiting the Victoria and Albert Museum alone or in a group by main subject area studied

Visiting:	Visual arts	Other 'arts' subjects	Science	Other or general studies	All subjects
	%	%	%	%	%
Alone	34	25	23	34	33
With friends or family	41	56	71	49	51
With a school party	8	10	6	4	5
With another type of organised party	16	9	0	12	12
Weighted base	*70*	*60*	*30*	*70*	*170*

Note that some individuals had studied more than one subject area.

Source of table: weighted results from the count-based sample of museum leavers.

Table 3.21 Percentages of full-time students aged 17 and over visiting the Science Museum alone or in a group, by main subject area studied

Visiting:	Visual arts	Other 'arts' subjects	Science	Other or general studies	All subjects
	%	%	%	%	%
Alone	—	—	22	20	20
With friends or family	—	—	54	50	58
With a school party	—	—	4	11	4
With another type of organised party	—	—	20	19	18
Weighted base	*10*	*20*	*50*	*30*	*90*

Note that some individuals had studied more than one subject area.

Source of table: weighted results from the count-based sample of museum leavers.

with their studies may be about twice as high as their answers to the question on the main purpose of their visit imply. Eighteen per cent of the science students in the quota sample (the same proportion as in the count-linked sample) cited their studies as a main reason for visiting. However, Table 3.25 shows that when they were asked directly twice that number agreed that the visit would help with their studies. Table 3.24 gives equivalent quota-based figures for the Victoria and Albert Museum. The figure for the proportion of students of visual arts who gave their studies as a main reason for their visit is, at 31 per cent, a good deal higher than the figure derived from the representative count-linked sample. This suggests that 67 per cent, the table's figure for the proportion of art students who thought the visit would help their studies, may also be too high. The important point, which need not have been affected by the overall bias of the table, is that for each art student who gave his or her studies as a main reason there was at least one other who thought the visit would help. Applying this proportion to the count-based figure in Table 3.22 suggest that a third or more of the museum's non-party art student visitors thought that the visit would help with their studies.

Table 3.26 shows how the students, whatever their subjects, thought their visits to the two London museums would benefit their studies. Nearly all referred

Table 3.22 Percentages of full-time students aged 17 and over citing studies as a main reason for visiting the Victoria and Albert Museum by main subject area studied

	Visual arts	Other 'arts' subjects	Science	Other or general studies	All subjects
All students	23%	13%	0%	12%	12%
Weighted base	*70*	*60*	*30*	*70*	*170*
Students visiting alone or with friends or family	17%	9%	0%	7%	8%
Weighted base	*50*	*50*	*20*	*60*	*140*

Note that some individuals had studied more than one subject area.

Source of table: weighted results from count-based sample of museum leavers.

Table 3.23 Percentages of full-time students aged 17 and over citing studies as a main reason for visiting the Science Museum, by main subject area studied

	Visual arts	Other 'arts' subjects	Science	Other or general studies	All subjects
All students	—	—	19%	11%	13%
Weighted base	*10*	*20*	*50*	*80*	*90*
Students visiting alone or with friends or family	—	—	18%	0%	9%
Weighted base	*10*	*10*	*40*	*20*	*70*

Note that some individuals had studied more than one subject area.

Source of table: weighted results from count-based sample of museum leavers.

Table 3.24 Importance of studies in visits to the Victoria and Albert Museum by full-time students aged 17 and over, by main subject areas studied

	Visual arts	Other 'arts' subjects	Science	Other or general studies	All subjects
Studies cited as a main reason for visiting the museum	31%	16%	7%	6%	11%
Thought visiting the museum would help with studies	67%	29%	13%	17%	29%
Base	*60*	*70*	*45*	*72*	*178*

Note that some individuals had studied more than one subject area.

Source of table: the quota samples of museum leavers.

Table 3.25 Importance of studies in visits to the Science Museum by full-time students aged 17 and over, by main subject areas studied

	Visual arts	Other 'arts' subjects	Science	Other or general studies	All subjects
Studies cited as a main reason for visiting the museum	22%	8%	18%	9%	12%
Thought visiting the museum would help with studies	53%	21%	35%	16%	24%
Base	*19*	*48*	*78*	*87*	*172*

Note that some individuals had studied more than one subject area.

Source of table: the quota samples of museum leavers.

Table 3.26   How full-time students aged 17 and over thought visiting the Victoria and Albert Museum or the Science Museum would help with their studies

Visit would:	Victoria and Albert Museum	Science Museum
	%	%
Broaden understanding	84	95
Provide practical ideas	14	5
Help in other ways	10	2
Base	*51*	*42*

The percentages add to over 100% as people could give more than one answer.

*Students who thought visiting the museum would help with their studies.

Source of table: the quota samples of museum leavers.

to the way the visit had extended their understanding. Few students at either museum had spotted any ideas which they intended to put into practice themselves.

4 Interviewing visitors to selected galleries

This is the first of four chapters based on interviews with people leaving particular galleries. The next three chapters will deal in order with visitors' reactions to the subject matter and contents of each gallery, their views about the methods of presentation used and finally what differences there are between the reactions of different categories of visitor. This chapter sets the scene for the following chapters by describing each of the ten galleries which were chosen for this aspect of the survey. It also gives the proportions of visitors who stopped to look at particular exhibits in each gallery.

Before going on, a word about the sample is in order. The samples were based on quotas in the way described in Chapter 1 and Appendix 1. This means that the samples selected reflect the need to have balanced quotas—the need for sufficient numbers of solitary visitors, people in different age groups, women as well as men—in order to analyse the effect of these variables on viewing patterns. As a result the composition of the samples do not reflect the actual composition of each gallery's visiting public. The kinds of people included in the gallery quotas were the same as those included in the quotas of people leaving each museum except that the gallery quotas included some children aged 11 and over who were on a school visit to the museum concerned.

Visitors were asked at the start of the interview:

"Did you stop to look at any of the things in (the gallery concerned) or did you walk straight through without stopping to look at any of the exhibits?"

If the visitor said that he or she had walked straight through, the interview was closed and the individual was not included in the quota. However any visitor who said that he or she had 'stopped to look' at something was considered eligible.

The phrase 'stopped to look' will occur in various contexts in the next few chapters, since we used it when asking members of the gallery samples which exhibits they had given attention to and when asking members of the quota samples of museum leavers which galleries they had visited. The phrase was chosen because we felt it implied a definite act of attention on the part of the visitor which went somewhat beyond just glancing round at the exhibits while strolling through the gallery.

In order to discover which exhibits had attracted each visitor's attention the interviewers showed people interviewed leaving each gallery a detailed plan of the gallery concerned and asked *"Can you show me what*

you stopped to look at during your visit today?". The main question was followed by a number of further queries to minimise omissions. It is important to keep in mind the wording of the question. The items mentioned in answer to it would be ones to which the visitor definitely directed his or her attention, sufficiently to stop in order to look at them. This does not imply that the visitor studied each of the items in detail. But on the other hand exhibits which a visitor merely glanced at while strolling through the gallery would not be included, even though these exhibits might contribute to the general impression the visitor took away.

Shortly after identifying the items which they had stopped to look at visitors were asked *"About how long have you spent in* [the gallery concerned] *today?"* Their answers provided the basis for the results concerning the length of people's visits to each gallery, discussed in this chapter and in Chapter 5.

Some explanation is needed of the reasons that led us to choose the particular galleries we did for detailed study. Though we used no formal criteria to select them, the ten galleries concerned—four each at the Victoria and Albert Museum and the Science Museum and two at the National Railway Museum—were chosen, in consultation with the staff of each museum, because they were felt to offer substantial contrasts in terms of both subject matter and presentation and would therefore enable us to investigate the ways in which visitors responded to different types of gallery design. The descriptions, which follow, of the ten different galleries are intended to give the reader an idea of the contrasts between them as well as of the ways in which they resemble each other.

a. The Victoria and Albert Museum
Our plan of the Victoria and Albert Museum (see pages 2–3) shows over 50 separate galleries each apparently occupying a clearly defined area within the museum. In fact the arrangement is not as rigid as the plan may suggest. So, before going on to discuss individual galleries, it is worth outlining the general principles which underlie the division of the museum's building and collection into different galleries. There are two sets of galleries, both open to the public: *primary galleries* which group together diverse works belonging to the same area or historical period, and *study collections* grouping together examples of the same art form; for instance textiles, metalwork and sculpture are the subjects of different study collections. Primary galleries dealing with related geographical regions or historical

TUDOR GALLERY

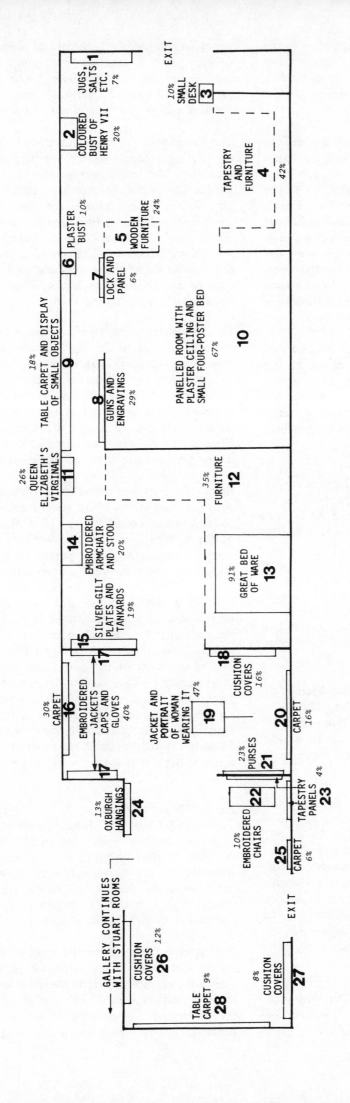

The exhibit numbers in heavy print were assigned to ease recording in the interviews.
The italicised percentages show the proportion of informants who had stopped to look at each exhibit.

periods are frequently located next to each other in the building, as are study galleries covering similar types of work, so that the division between galleries is often more a matter of gradation than of the clearly marked boundaries that appear on our plan.

Since the exhibition spaces devoted to different subjects shade into one another it is frequently no more than a matter of convenience whether one refers to a particular area as one gallery or as two. An example of this is the gallery marked on our museum plan as 'British Art: Tudor and Early Stuart.' It was decided to include this in the set of four galleries to be studied in detail. However since the gallery defined in this way was rather large, containing a number of different rooms, the gallery interviews dealt only with visits to the rooms relating to the Tudor period—which our interviewers were instructed to refer to as the 'Tudor gallery'.

The Tudor gallery is a primary gallery as were two of the other galleries selected for gallery interviewing: the Continental Seventeenth Century gallery and the Art of China and Japan gallery. The fourth gallery selected was the British Sculpture gallery. This gallery differed from the other three in that it contained only sculpture.

Tudor

The Tudor gallery has been renovated since the survey was carried out. The detailed plan (see page 31) shows how the gallery was set out at the time of the survey. It consisted of a single wide corridor divided into five small rooms. The interviewer was stationed in the room containing items 22 to 28 and contacted visitors as they left the gallery either by the door between items 25 and 27 or by the exit on the other side of the room which led to the Stuart rooms. The great majority of the people interviewed therefore will have entered the gallery at the opposite end, starting in the room containing the bust of Henry VII.

The plan shows what proportion of our sample of visitors stopped to look at each exhibit. The most viewed exhibits were the Great Bed of Ware (item 13 on the plan) and the panelled room (item 10 on the plan) mentioned by 91 per cent and 67 per cent respectively of the visitors.

Other exhibits which had attracted a fair degree of attention were an embroidered jacket placed next to a portrait of its owner (item 19, looked at by 47 per cent), two cases containing embroidered jackets, caps and gloves (item 17, 40 per cent) and a tapestry and various items of furniture that were on the visitors' left as they first entered the gallery (item 4, 42 per cent). Textiles, furniture and various small objects made up the remaining exhibits. In the interests of conserving the exhibits the gallery was only dimly lit.

Continental Seventeenth Century

The Continental Seventeenth Century gallery also consists of a long corridor divided into a number of separate rooms. It is entered through self-closing glass doors placed there to control the atmosphere in the gallery, which is carefully maintained at a constant temperature and humidity level. Much of the floor is carpeted and careful lighting brings out the highly finished quality of all the objects on show.

The interviews took place outside the gallery itself, in the High Renaissance gallery (see the plan of the whole museum). The interviewer contacted visitors as they left the Continental Seventeenth Century gallery by the glass door and stairs leading from the room containing small scale furnishings (item 26) at the right-hand end of the detailed plan (see page 33). Nearly all the people interviewed would have entered the gallery either by the circular room at the left-hand end of the gallery plan or by the staircase leading down to the room containing items 12 to 15.

Of the visitors interviewed, 42 per cent remembered stopping to look at the highly decorated silver clock (item 3 on the plan) in the room at the left-hand end of the gallery. The sculpture of the Nativity (item 2, looked at by 34 per cent), the tiles in the second room (item 4, 33 per cent), the clothing in the next room (item 10, 34 per cent) and the dark panelled room (item 11, 39 per cent) were other items in the first part of the gallery which attracted particular attention.

Some of the visitors would have entered the gallery in the middle by means of the stairs leading down to the room containing exhibits 12 to 15. Thirty-nine per cent of the visitors looked at the furniture in that room (item 12) and slightly more paused to look at the wax sculptures, which included a gruesome tableau on the theme of Time and Death, on either side of the narrow corridor leading into the next room (item 16). The pottery and metalwork displayed in glass cases set into the sides of this octagonal room (item 17, 46 per cent) and the gold and silver work displayed in cases set into one of the walls of the next room (item 20, 44 per cent) appear to have caught the eye of many of the visitors. The three groups of exhibits which most visitors reported looking at were the sculpture (item 22, 53 per cent), the furniture in the next room (item 24, 53 per cent) and the small scale furnishings in the last room of the gallery (item 26, 64 per cent).

Art of China and Japan

The Art of China and Japan gallery is housed in a large, high-ceilinged room and an adjoining stretch of corridor. As the detailed plan (see page 34) shows, the interviews for this survey were only concerned with the exhibition in the main room. The interviewer contacted visitors who were leaving the gallery by the exit that leads into the Early Medieval gallery.

The exhibits which attracted most notice were the throne and screen which dominate the centre of the gallery as one looks towards the left-hand of our plan (item 14, looked at by 57 per cent of the visitors) and the Ch'ien-lung throne opposite the exit to the Early Medieval gallery. Other particularly popular exhibits

CONTINENTAL SEVENTEENTH CENTURY GALLERY

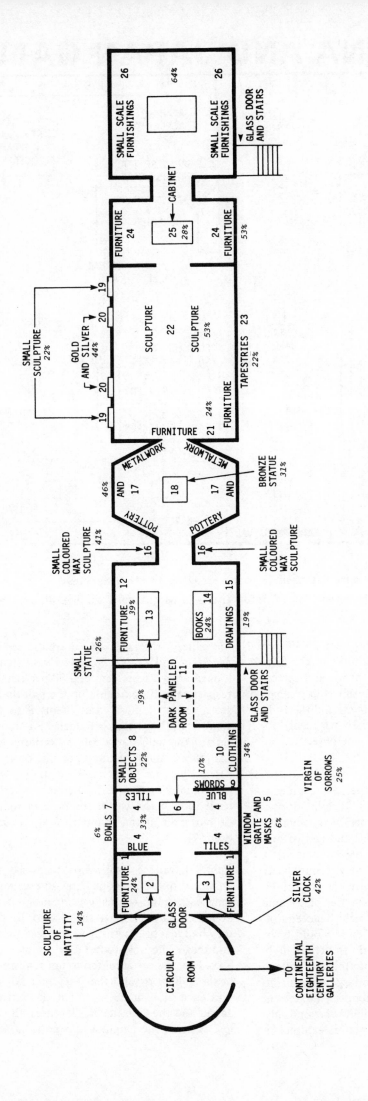

The exhibit numbers in heavy print were assigned to ease recording in the interviews.

The italicised percentages show the proportion of informants who had stopped to look at each exhibit.

ART OF CHINA AND JAPAN GALLERY

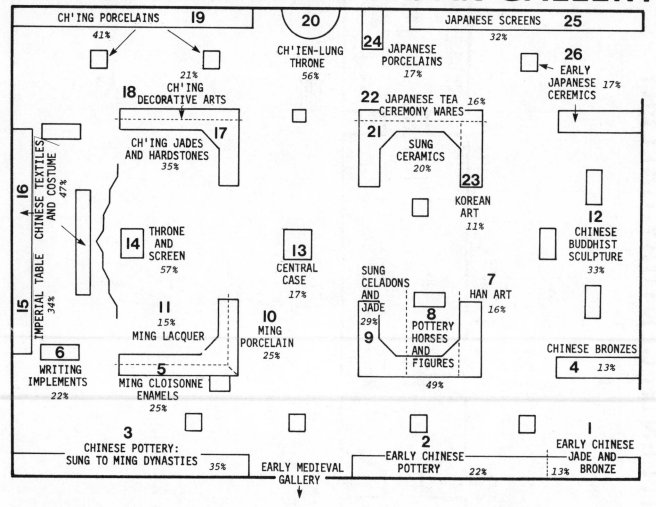

The exhibit numbers in heavy print were assigned to ease recording in the interviews.

The italicised percentages show the proportion of informants who had stopped to look at each exhibit.

were the pottery horses and figures (item 8, 49 per cent) and the textile and costume exhibits at the left-hand end of the gallery behind the throne and screen (item 16, 47 per cent). The exhibits in the gallery range in size from the large thrones and screens and the full-size Buddhist sculptures (item 12, 33 per cent) to pottery and small, intricately carved jade objects.

British Sculpture

The British Sculpture gallery (see page 35) occupies a large hall with high, green-grey walls. It is lit, during the day, by natural light which comes in through its glass roof. It is closed off at the left-hand end of our plan by a massive carved stone screen (or 'rood loft') from a Flemish Church. The interviewer contacted visitors as they left the gallery between the sculptures of *Valour and Cowardice* (item 1 on the plan) and *Truth and Falsehood* (item 5) at the right-hand end of our plan. Most of the visitors interviewed would have entered the gallery at the opposite end through an arch at the base of the rood loft. On entering the gallery they would see in front of them the large monument to Sir Henry Moyle and his wife (item 28, looked at by 67 per cent of the visitors). Immediately beyond this they would come upon the two most viewed exhibits in

the gallery, the large and dramatic statues of *Raving Madness* and *Melancholy Madness* (items 24 and 25 looked at by 71 per cent and 70 per cent of the visitors respectively). Looking on up the hall the visitors would see a series of glass cases (items 8 to 19) containing small sculptures, sketch models and so on. Along each side of the hall they would see a large number of full sculptures, many of them portrait busts (items 30 and 31) as well as a number of larger works marked individually on the plan. The place where the interview itself took place was at the bottom right-hand corner of our plan in the shadow of *St Michael and Satan* (item 7).

Tables 4.1 and 4.2 show how members of the samples answered questions about how interesting or enjoyable they had found the gallery in question. They divided about equally between those who found the gallery they had just been visiting very interesting or enjoyable, and those who only found it fairly interesting or enjoyable. Few visitors admitted to not really being interested in the gallery or said that they had not really enjoyed visiting it. It can be seen that the Art of China and Japan and the Continental Seventeenth Century galleries earned the highest proportion of favourable

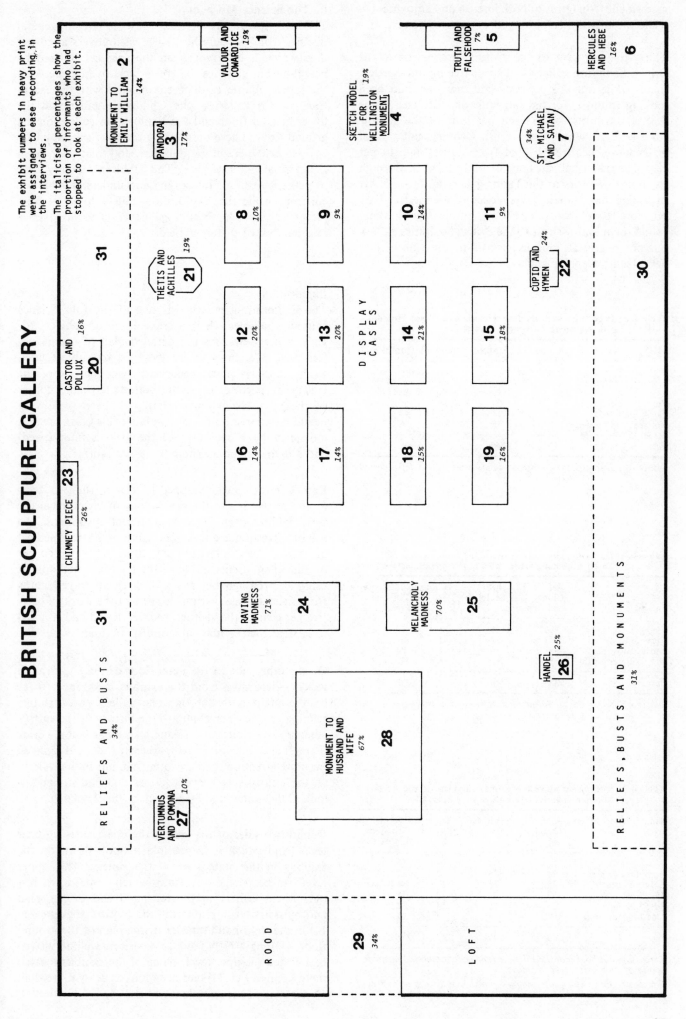

BRITISH SCULPTURE GALLERY

The exhibit numbers in heavy print were assigned to ease recording, in the interviews.

The italicised percentages show the proportion of informants who had stopped to look at each exhibit.

VALOUR AND COWARDICE *19%* **1**

MONUMENT TO EMILY WILLIAM **2** *14%*

PANDORA **3** *17%*

SKETCH MODEL FOR WELLINGTON MONUMENT *19%* **4**

TRUTH AND FALSEHOOD *7%* **5**

HERCULES AND HEBE *16%* **6**

ST. MICHAEL AND SATAN *34%* **7**

8 *10%*

9 *9%*

10 *14%*

11 *9%*

THETIS AND ACHILLES **21** *19%*

CUPID AND HYMEN *24%* **22**

DISPLAY CASES

12 *20%*

13 *20%*

14 *21%*

15 *18%*

16 *14%*

17 *14%*

18 *15%*

19 *16%*

CASTOR AND POLLUX *16%* **20**

CHIMNEY PIECE **23** *26%*

RELIEFS AND BUSTS **31** *34%*

30

RELIEFS, BUSTS AND MONUMENTS *31%*

RAVING MADNESS *71%* **24**

MELANCHOLY MADNESS *70%* **25**

HANDEL **26** *25%*

VERTUMNUS AND POMONA **27** *10%*

MONUMENT TO HUSBAND AND WIFE *67%* **28**

R O O D

29 *34%*

L O F T

35

assessments—in terms of both interest and enjoyment—while the British Sculpture gallery received fewest.

Thirty-five per cent of the visitors interviewed in the British Sculpture gallery had spent five minutes or less there, while only 16 per cent said that they had been looking round it for half an hour or more (see Table 4.3). By contrast only seven per cent of the sample for the Continental Seventeenth Century gallery had walked briskly through it in five minutes and 44 per cent had taken half an hour or more to walk through its rooms. Visitors to the Tudor gallery and to the Art of China and Japan gallery occupy an intermediate position: they spent more time on average looking round them than visitors to the British Sculpture gallery but not nearly as long as visitors to the Continental Seventeenth Century gallery.

Table 4.1 Interest levels in the four Victoria and Albert Museum galleries where there were interviews

	Art of China and Japan	British Sculpture	Tudor Art	Continental 17th Century Art
	%	%	%	%
Very interested	52	37	44	50
Fairly interested	43	53	47	44
Not really interested	5	10	9	6
Base	232	240	251	236

Source of table: the quota samples interviewed as they left selected galleries.

Table 4.2 How enjoyable visitors found the four Victoria and Albert Museum galleries where there were interviews

	Art of China and Japan	British Sculpture	Tudor Art	Continental 17th Century Art
	%	%	%	%
Very enjoyable	51	32	41	46
Fairly enjoyable	44	56	54	48
Not really enjoyable	5	12	5	6
Base	232	240	251	236

Source of table: the quota samples interviewed as they left selected galleries.

Table 4.3 Time spent by visitors in the four Victoria and Albert Museum galleries where there were interviews

Time spent in gallery	Art of China and Japan	British Sculpture	Tudor Art	Continental 17th Century Art
	%	%	%	%
5 minutes or less	21	35	27	7
6–14 minutes	28	**28**	**27**	14
15–19 minutes	**33**	22	30	**35**
30–59 minutes	13	12	14	32
One hour or more	6	4	2	12
Base	232	240	251	236

Percentages in bold indicate the time range containing the median length of time spent in each gallery.

Source of table: the quota samples interviewed as they left selected galleries.

b. The Science Museum

The galleries at the Science Museum are in general distinct from one another, since each covers a subject area which is different from the subjects covered by neighbouring galleries. There is little overlap between the subject matter of different galleries with the exception of two galleries—the Children's gallery on the lower ground floor and the Exploration gallery on the ground floor. These both take a range of topics, partly covered elsewhere in the museum, and treat them from a particular point of view. The Children's gallery aims to make its exhibits interesting and understandable for children, while the Exploration gallery looks at how recent advances in technology have led to discoveries in a number of different fields.

Exploration
The Exploration gallery was one of the four Science Museum galleries which were selected as subjects for gallery interviews. As the detailed plan of the gallery (see page 37) shows, it is arranged by subject area. Each section is partly separated from the rest of the gallery by partitions. At the bottom left-hand corner of the plan are exhibits relating to deep sea diving including a large model of a submersible craft (item 1, looked at by 47 per cent of the visitors interviewed) and a diving chamber (item 3, 58 per cent).

Rather fewer visitors stopped to look at the two items (4 and 5) relating to the psychology of vision or at the three exhibits (items 6 to 8) of recent advances in the use of ultrasound, X-rays and fibre optics for medical diagnosis. The exhibits in the bottom right-hand corner of our plan (items 17 and 18) concern methods of using electromagnetic rays with different wave lengths from light, to 'see' various aspects of the world. Forty-five per cent of the visitors we interviewed had stopped to look at the thermal camera display (item 17).

On the other side of the main floor of the Exploration gallery—separated from the exhibits mentioned so far by the main path through the gallery—are exhibits relating to space travel and to the history of the earth's climate. The model of an American lunar landing craft (item 9) and the *Apollo 10* spacecraft (item 10) received far more attention than any other exhibits in the gallery. Of the visitors interviewed, 85 per cent had stopped to look at the lander and 86 per cent at the *Apollo 10*.

Round two sides of the gallery there is a narrow upper level from which it is possible to look down on the exhibits in the main part of the gallery. This upper level can be reached by stairs which come down just outside the gallery at the left-hand end of our plan and by stairs leading up from the *Skylark* rocket (item 11) in the right-hand quarter of the plan of the ground floor. Comparatively few of the visitors we interviewed had visited the uper level. None of the exhibits located there (items 19 to 24) had been looked at by more than 14 per cent of the sample members.

EXPLORATION GALLERY

The exhibit numbers in heavy print were assigned to ease recording in the interviews.

The italicised percentages show the proportion of informants who had stopped to look at each exhibit.

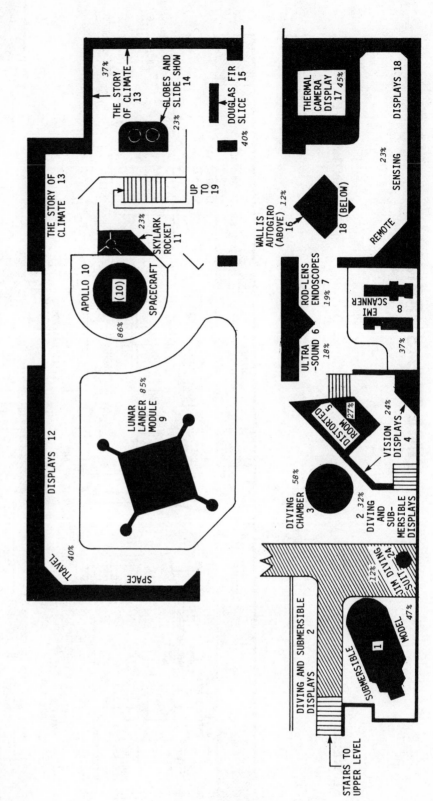

TIME MEASUREMENT GALLERY

CENTRAL WELL

CENTRAL WELL

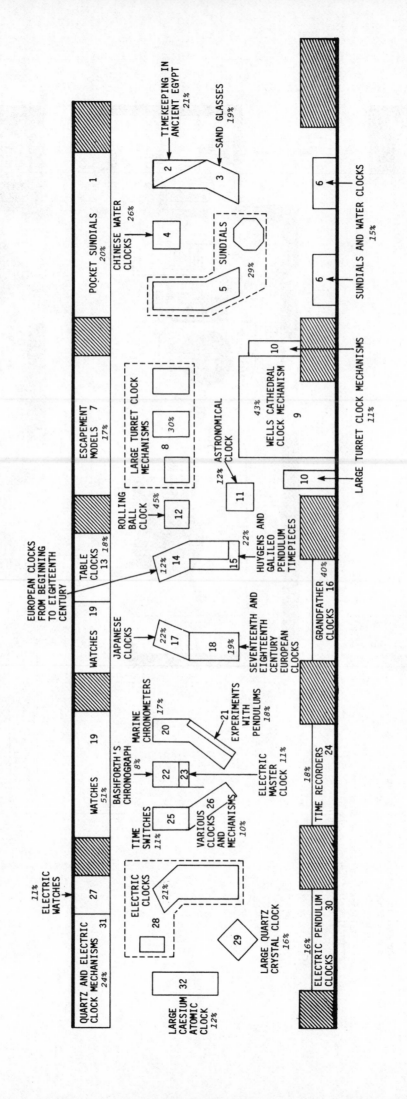

QUARTZ AND ELECTRIC CLOCK MECHANISMS 31	*24%*
11% ELECTRIC WATCHES 27	
WATCHES 51% 19	
WATCHES 19	
ESCAPEMENT MODELS 7 *17%*	
POCKET SUNDIALS 20% 1	

ELECTRIC CLOCKS 28 *21%*

LARGE QUARTZ CRYSTAL CLOCK 29 *16%*

LARGE CAESIUM ATOMIC CLOCK 32 *12%*

TIME SWITCHES 25 *11%*

VARIOUS CLOCKS AND MECHANISMS 26 *10%*

BASHFORTH'S CHRONOGRAPH *8%* 22 23

MARINE CHRONOMETERS 20 *17%*

EXPERIMENTS WITH PENDULUMS 21 *18%*

ELECTRIC MASTER CLOCK *11%*

EUROPEAN CLOCKS FROM BEGINNING TO EIGHTEENTH CENTURY

TABLE CLOCKS 13 *18%*

14 *12%*

HUYGENS AND GALILEO PENDULUM TIMEPIECES 15 *22%*

JAPANESE CLOCKS 17 *22%*

18 *19%*

SEVENTEENTH AND EIGHTEENTH CENTURY EUROPEAN CLOCKS

ROLLING BALL CLOCK 12 *45%*

LARGE TURRET CLOCK MECHANISMS 8 *30%*

ASTRONOMICAL CLOCK 11 *12%*

CHINESE WATER CLOCKS *26%* 4

SUNDIALS 5 *29%*

TIMEKEEPING IN ANCIENT EGYPT 2 *21%*

SAND GLASSES 3 *19%*

ELECTRIC PENDULUM CLOCKS 30 *16%*

TIME RECORDERS 24 *18%*

GRANDFATHER CLOCKS 16 *40%*

LARGE TURRET CLOCK MECHANISMS 10 *11%*

WELLS CATHEDRAL CLOCK MECHANISM 9 *43%*

SUNDIALS AND WATER CLOCKS 6 *15%*

6

The exhibit numbers in heavy print were assigned to ease recording in the interviews.
The italicised percentages show the proportion of informants who had stopped to look at each exhibit.

The interviewer contacted visitors as they left the gallery between the Douglas fir slice (item 15) and the back of the thermal camera display (item 17) at the right-hand end of the plan, and interviewed them just outside the gallery.

The gallery is dark except for bright lighting directed at the often colourful exhibits. Black partitions rise from the darkly carpeted floor to the ceiling of the gallery 20 feet or so above the visitor's heads. Visitors entering the gallery are greeted by the sound of loudspeaker and television commentaries on the exhibits. The gallery offers a range of audio-visual information sources, including telephones and slide shows as well as television sets and loudspeakers, alongside more traditional labels and written information.

The other Science Museum galleries selected for special attention in this survey were the Time Measurement gallery on the first floor of the museum, the Printing and Paper gallery on the second floor, and the Aeronautics gallery on the third floor.

Time Measurement

The Time Measurement gallery (page 38) occupies a narrow balcony on one side of a central well. Visitors can look down onto the transport galleries on the ground floor. The artificial lighting in the gallery is supplemented by windows set high up in the wall opposite the central well. The presentation of the gallery is fairly traditional, most of the exhibits are in glass cases and information is provided solely in the form of labels and written explanations. On the whole the individual exhibits are small. The boundary between the Time Measurement gallery and the Astronomy gallery is marked by a change in the size of exhibits and in the prevalence of glass cases. However there is little difference between the general appearance of the Time Measurement gallery and that of its other neighbour, the Meteorology gallery.

One way in which the Time Measurement gallery does differ from its neighbours is that many of its exhibits are in operation. Every quarter of an hour the gallery is shaken by the chimes of the very large Wells Cathedral clock mechanism (item 9 on the detailed plan of the gallery). Forty-three per cent of the visitors interviewed said that they had stopped to look at the mechanism while they were in the gallery. Other exhibits which attracted a good deal of attention were the rolling ball clock (item 12, looked at by 45 per cent of the visitors), the grandfather clocks (item 16, 40 per cent) and the cases of watches by the central well (item 19, 51 per cent). The interviewer was stationed at the left-hand end of the gallery as shown on the plan, so that most of the people interviewed would have walked through the gallery in chronological order. The first exhibits they would have seen were devoted to early methods of time measurement using sundials, sand glasses and water clocks (items 2 to 6) while the last ones they would have passed were of various modern technical and scientific uses of time measurement (items

23 to 26), electric clocks (items 27 to 31) and the caesium atomic clock (item 32). The exhibits in the centre of the gallery (items 7 to 22) deal with the development of weight and spring driven timepieces.

Printing and Paper

The decor of the Printing and Paper gallery (page 40) is predominantly green, white and red. The gallery is brightly lit, entirely by artificial light. Loudspeakers provide commentaries on various exhibits, sometimes supplemented by sequences of slides shown on small screens. Further sound is provided by the whirring and chattering of typesetting and printing machines (items 8 and 17 on the detailed plan) one of each is generally in operation. Written information is provided by commentaries in large print, sometimes accompanied by pictures, on walls and partitions, as well as along the railing separating the visitor from the machines set out on either side of the main part of the gallery (items 8 and 17 again). The name of each of these machines is given on a piece of card, about three or four feet square hanging from a sort of flagpole set into the wall behind and above the machine in question.

The interviewer contacted visitors as they left the gallery at the right-hand end of our plan, by items 9, 10 and 17. The interview took place just outside the gallery, but by a glass partition through which it could be seen.

The gallery's two other entrances from the rest of the second floor are shown on the left-hand end of the plan. One is by the earliest types of hand press (item 2) and the display relating to signs, writing and early printing (item 1), the other is through the area reserved for topical displays at the bottom left-hand corner of the plan. A fourth way of entering and leaving the gallery is provided by the museum's lifts which open into the Printing and Paper gallery, opposite item 1 on the plan.

The Printing and Paper gallery, like the Exploration gallery, has a small upper floor from which visitors can look down at the exhibits on the main floor. This upper level may be reached by steps leading up from the left-hand end of the display of typesetting machines. A further set of steps connects the upper level to the main floor just next to the exit at the lower left-hand corner of the plan.

Each of the main series of machines set out on either side of the main floor had been looked at by two thirds of the visitors whom we interviewed. Other exhibits which attacted a good deal of attention were the model of an eighteenth century printer's shop (item 3, looked at by 40 per cent of the visitors), the various hand presses (item 2, 47 per cent) and the display relating to paper and board making. This display, which was enclosed within six or seven foot high circular partitions, is shown towards the right of the plan as items 11 and 12, looked at by 46 per cent and 45 per cent, respectively of the visitors. Comparatively few of the

PRINTING AND PAPER GALLERY

The exhibit numbers in heavy print were assigned to ease recording in the interviews.

The italicised percentages show the proportion of informants who had stopped to look at each exhibit.

Plan of Upper Level

DOWN TO 7 & 8

22

TYPEWRITERS *18%*

22

WRITING IMPLEMENTS *7%*

21 **21**

DUPLICATORS *13%*

22

20

20

CINEMA **19** *2%*

NEWSPAPERS **18** *13%*

DOWN TO 5

Plan of Main Floor

SIGNS, WRITING AND EARLY PRINTING **1** *7%*

HAND PRESSES **2** *47%*

STAIRS TO UPPER LEVEL

MECHANICAL TYPESETTING **8** *67%*

COMPOSING TYPE BY HAND **7** *15%*

PHOTOSETTING **14** *30%*

LETTERS **9** *5%*

HAND PRESSES **2**

PRINTING PICTURES **15** *18%*

COLOUR PRINTING **16** *23%*

PHOTOSETTING **14**

HAND PRESSES **2**

18th CENTURY PRINTER'S SHOP **3** *40%*

LIFTS

STAIRS TO UPPER LEVEL

TYPE FOUNDING **4** *9%*

AREA FOR TOPICAL DISPLAYS **5** *5%*

PAPER AND BOARD MAKING **11** *46%*

MODELS OF PAPER AND BOARD MAKING MACHINES **12** *45%*

PAPER AND BOARD MAKING **11**

SPECIMENS OF PRINTED MATTER **13** *14%*

FINISHING PROCESSES BOOKBINDING **9** *26%*

PRINTING INK **10** *19%*

PRINTING MACHINES 19th AND 20th CENTURY **17** *69%*

40

visitors whom we interviewed had looked at the exhibits on the upper level, the typewriters (item 22, 18 per cent) having attracted most attention. Only two per cent of the visitors we spoke to had visited the gallery's small cinema as it was closed nearly all the time we were interviewing.

Aeronautics

The Aeronautics gallery was the largest of the four Science Museum galleries selected for special study. It is a long hall with a high roof curving down towards each side to meet slightly outward-sloping walls of glass, about ten feet high, through which daylight enters. At one end (the left-hand end on the detailed plan on page 42) the gallery narrows and leads out to a spiral staircase and an escalator beyond which are the other galleries on the third floor. Visitors sitting in the tea room (as the museum's restaurant is known) at the other end of the gallery can look out through large glass windows at the contents of the hall. Visitors leaving the gallery at the tea room end, whether for the tea room or for lower floors, go out through a corridor to one side of the tea room (this is shown most clearly on the map of the whole museum).

The space above the visitors' heads is thronged with aircraft suspended from the ceiling. In order to see the aircraft at eye-level visitors can use the raised walkway which runs the length of the gallery. By lifting telephone receivers placed along the walkway the visitors can hear commentaries on the aircraft which are visible from the point concerned. Visitors walking on the main floor of the gallery can look into cockpits from various types of aircraft, as well as inspecting numerous aircraft engines and looking at many cases of exhibits illustrating the history of flight. The gallery as a whole is grey-blue in colour. The ceiling is painted dark blue while the floor and many of the exhibits are grey. The showcases are painted light blue or grey.

The aircraft which had attracted most attention from the sample of visitors whom we interviewed were the *Spitfire* (item 36, which 63 per cent of the visitors had stopped to look at) and the *Hurricane* (item 34, 51 per cent) towards the right-hand end of the detailed plan; the *Vickers Vimy* in which Alcock and Brown flew the Atlantic (item 28, 45 per cent) in the centre of the gallery; and a copy of the aircraft in which the Wright brothers first flew (item 16, 39 per cent) at the left-hand end. Turning to the exhibits mounted on the floor of the gallery, the cockpits (item 10) were looked at by 43 per cent of our sample of visitors. Forty per cent had looked at one or more of the cases containing aircraft models (item 8). The exhibits of aircraft engines also attracted a good deal of attention (38 per cent and 39 per cent respectively having stopped to look at one or more of the exhibits grouped on the plan as items 5 and 12).

Visitors were contacted as they left the gallery at the tea room end (at the right-hand end on the detailed

plan) and interviewed either in the corridor leading from it or just inside the gallery itself.

Tables 4.4 and 4.5 show how interesting and how enjoyable the sample of visitors leaving each gallery reported it to be. Few reported that the gallery they were leaving was not really interesting or enjoyable. However there is a marked difference between the galleries in the proportions of visitors who gave strongly positive reactions. The Aeronautics gallery and the Exploration gallery are both given the highest of the three possible ratings for interest and enjoyment far more often than are the other two galleries. The differences in interest levels are matched by differences in the length of time visitors spent in each gallery. Forty-three per cent of the sample interviewed in the Aeronautics gallery and 51 per cent of the visitors contacted as they left the Exploration gallery had spent over half an hour in the gallery concerned. Comparable figures for the Printing and Paper gallery and the Time Measurement gallery being ten per cent and 16 per cent respectively (see Table 4.6). These differences in the length of gallery visits might of course be due in part to factors other than interest levels, for instance to differences in the sizes of the galleries concerned. One of the topics examined in Chapter 5 will be the relationship between visitors' interest levels and the time they spend in any given gallery.

Table 4.4 Interest levels in the four Science Museum galleries where there were interviews

	Aero-nautics	Printing and Paper	Time Measurement	Exploration
	%	%	%	%
Very interested	61	32	33	64
Fairly interested	35	59	60	33
Not really interested	4	9	7	3
Base	*255*	*222*	*227*	*243*

Source of table: the quota samples interviewed as they left selected galleries.

Table 4.5 How enjoyable visitors found the four Science Museum galleries where there were interviews

	Aero-nautics	Printing and Paper	Time Measurement	Exploration
	%	%	%	%
Very enjoyable	53	30	29	52
Fairly enjoyable	43	63	64	46
Not really enjoyable	3	7	7	2
Base	*255*	*222*	*227*	*243*

Source of table: the quota samples interviewed as they left selected galleries.

AERONAUTICS GALLERY

The exhibit numbers in heavy print were assigned to ease recording in the interviews.

The italicised percentages show the proportion of informants who had stopped to look at each exhibit.

Suspended Aircraft

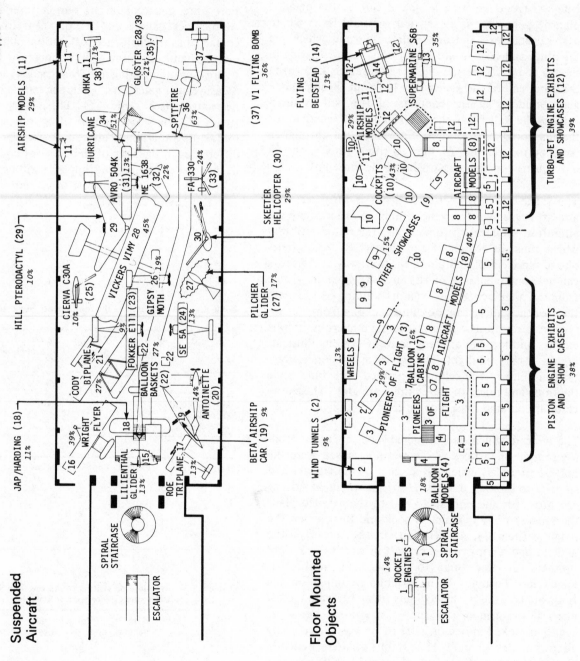

Floor Mounted Objects

42

Table 4.6 Time spent by visitors in the four Science Museum galleries where there were interviews

Time spent in gallery	Aero-nautics	Printing and Paper	Time Measurement	Exploration
	%	%	%	%
5 minutes or less	10	26	21	6
6–14 minutes	15	**35**	33	16
15–29 minutes	**33**	29	30	27
30–59 minutes	29	8	14	**36**
One hour or more	14	2	2	15
Base	255	222	227	243

The percentages in bold indicate the time range containing the median length of time spent in each gallery.
Source of table: the quota samples interviewed as they left selected galleries.

c. The National Railway Museum

Nearly all the exhibits in the National Railway Museum are to be found either on the floor of the museum's main hall or on the balcony overlooking it. We therefore selected the Main Hall and the Balcony as the two 'galleries' that would be studied in depth at the National Railway Museum.

Main Hall

The Main Hall of the National Railway Museum contains far more exhibition space than the rest of the museum put together. A general idea of the hall's layout is provided by the plan of the whole museum (see page 44) while the detailed plan on page 45 gives a fuller picture. The hall is dominated by locomotives and rolling stock ranged round two turntables set into the floor. The turntable at the left of our plans is mainly reserved for locomotives, while most of the tracks radiating from the right-hand turntable are occupied by carriages. Round the side of the hall are exhibits illustrating various aspects of the operation of railways, past and present, including stationary engines, models, and displays illustrating the operation of points and signals.

The label beside each locomotive or carriage gives its name and main specifications in bold print, followed by what is really a short essay on the vehicle's design and history. Raised platforms are provided beside some of the exhibits to enable the visitors to see the interior of the locomotives' cabs or the furnishings of carriages.

The chief way into the Main Hall is at the lower left-hand corner of our plans, down the steps from the main entrance. This is also the only route by which a visitor to the Main Hall can leave the museum. The garden to the right of the hall on our plan can only be reached through the hall itself. Visitors to the balcony overlooking the Main Hall can enter through the restaurant but can only leave by walking down one of the two sets of stairs to the floor of the Main Hall.

Most of our information about people's visits to the Main Hall was provided by people who were contacted as they left the hall by the entrance at the bottom left of our plan. They were interviewed, using the 'gallery' questionnaire, just inside the hall itself. However some useful information was also provided by the quota sample of visitors who were interviewed when they finally left the whole museum. The data provided by this second set of visitors showed that nearly all visitors walked all round the Main Hall—99 per cent had looked at the locomotives, 94 per cent at the coaches and 88 per cent had stopped to look at some of the exhibits at the side of the hall.

The exhibit which attracted the most attention was the record-breaking *Mallard* locomotive (item 6) which 82 per cent of the visitors interviewed inside the hall itself had stopped to look at. Sixty per cent had looked at item 13, a steam locomotive with its side cut away to illustrate how the inside of its boiler is arranged. Fifty-nine per cent stopped to look at the *Western Fusilier* (item 4, a large red diesel locomotive). Fifty-six per cent of the visitors had looked at the locomotive placed over an inspection pit (item 7), though it is unlikely that all these people actually went down into the inspection pit. The items of rolling stock which attracted most attention were the royal coaches (item 18). These were second only to *Mallard* in the proportion of visitors (72 per cent) who stopped to look at them.

Only part of the museum's collection of locomotives and rolling stock is on show in the Main Hall at any particular time. Locomotives and carriages inside the hall are regularly moved out in order to make room for others—a process which often entails shunting round some of the other vehicles remaining in the hall. This made it difficult to draw up an accurate plan of the hall. Where the description given by our original plan of the vehicle or vehicles occupying a particular space round either turntable proved to be misleading the description has been left off the plan shown in this report. Six items (10, 11, 12, 14, 22 and 23) are affected. No figures are given for the proportions of visitors who stopped to look at the vehicles shown in these positions.

Balcony

The other part of the museum selected for gallery interviews was the balcony overlooking the Main Hall. The interviewer contacted visitors as they left the balcony at the left-hand end as shown on the plan of the whole museum, and interviewed them on a landing half way down the stairs to the Main Hall.

Someone starting their visit at the opposite end of the gallery would walk along a long series of displays relating the history of railways and their evolving place in British life. As the detailed plan (page 46) shows, this portion of the gallery is divided into a series of alcoves by show-cases and screens projecting from either side. Each case, or set of cases, deals with a different aspect of the story, illustrating it with models, pictures, dioramas and in a few instances with a small silent screen showing slides. The cases contain a great deal of written material, organised into three kinds.

NATIONAL RAILWAY MUSEUM PLAN

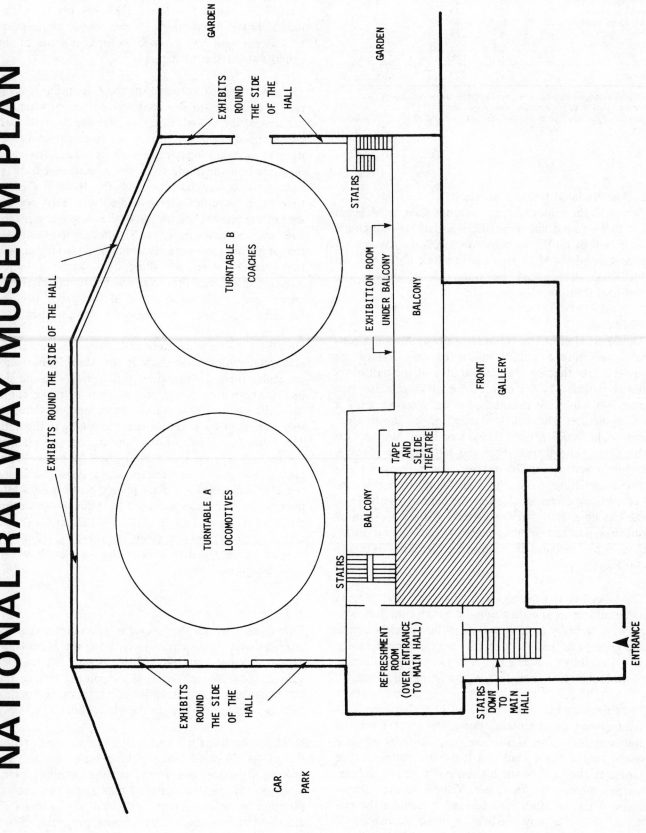

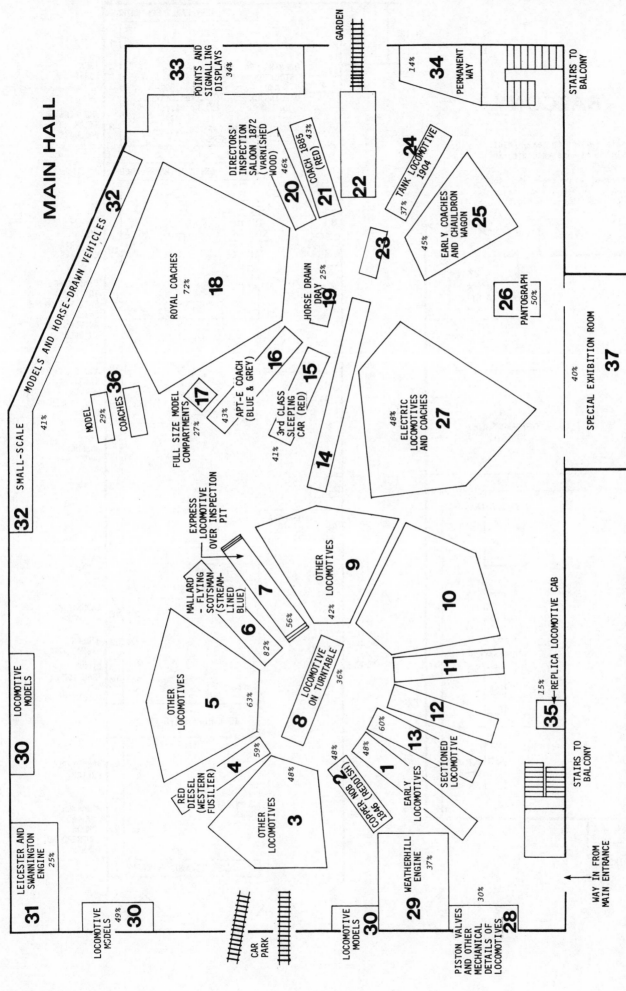

MAIN HALL

MODELS AND HORSE-DRAWN VEHICLES 32

SMALL-SCALE 32   41%

LOCOMOTIVE MODELS 30

LEICESTER AND SWANNINGTON ENGINE 31   25%

LOCOMOTIVE MODELS 30   49%

POINTS AND SIGNALLING DISPLAYS 33   34%

GARDEN

PERMANENT WAY 34   14%

STAIRS TO BALCONY

DIRECTORS' INSPECTION SALOON 1872 (VARNISHED WOOD) 20   46%

COACH 1885 (RED) 21   43%

22   23

TANK LOCOMOTIVE 1904 24   37%

EARLY COACHES AND CHAULDRON WAGON 25   45%

ROYAL COACHES 18   72%

MODEL 36   29%

COACHES

FULL SIZE MODEL COMPARTMENTS 17   27%

APT-E COACH (BLUE & GREY) 16   43%

3rd CLASS SLEEPING CAR (RED) 15   41%

HORSE DRAWN DRAY 19   25%

14

PANTOGRAPH 26   50%

ELECTRIC LOCOMOTIVES AND COACHES 27   48%

SPECIAL EXHIBITION ROOM 37   40%

EXPRESS LOCOMOTIVE OVER INSPECTION PIT

MALLARD – FLYING SCOTSMAN (STREAMLINED BLUE) 6   82%

7   56%

OTHER LOCOMOTIVES 5   63%

RED DIESEL (WESTERN FUSILIER) 4   59%

OTHER LOCOMOTIVES 3   48%

COPPER NOB (REDDISH) 1846 2   48%

LOCOMOTIVE ON TURNTABLE 8   36%

OTHER LOCOMOTIVES 9   42%

OTHER LOCOMOTIVES 10

11

12

SECTIONED LOCOMOTIVE

13   48%

EARLY LOCOMOTIVES 1   60%

REPLICA LOCOMOTIVE CAB 35   15%

STAIRS TO BALCONY

WAY IN FROM MAIN ENTRANCE

WEATHERHILL ENGINE 29   37%

LOCOMOTIVE MODELS 30

PISTON VALVES AND OTHER MECHANICAL DETAILS OF LOCOMOTIVES 28   30%

CAR PARK

The exhibit numbers in heavy print were assigned to ease recording in the interviews.

The italicised percentages show the proportion of informants who had stopped to look at each exhibit.

45

BALCONY

The exhibit numbers in heavy print were assigned to ease recording in the interviews.

The italicised percentages show the proportion of informants who had stopped to look at each exhibit.

STAIRS

53%

1 VERY EARLY BEGINNINGS

1

2 IRON HORSE *32%*

56%
OPENING OF STOCKTON AND DARLINGTON RAILWAY

3

3

1860 RAILWAY NAVVY

4

THE RAILWAY BUILDERS

31% **6**

5

THE FIRST STEAM RAILWAYS *41%*

5

43%

7 BRITAIN TRANSFORMED

29%

8

9
MODEL OF EUSTON STATION AS FIRST BUILT

44%

TICKETS *38%*

MID-VICTORIAN RAILWAYS

10

10 *48%*

11 *41%*

'FINISHING TOUCHES'- LATE VICTORIAN RAILWAY BRIDGES AND TUNNELS

WORKS PLATES AND MODELS **24** *37%*

24

12 *50%* LATE VICTORIAN AND EDWARDIAN RAILWAYS- WORKING CONDITIONS AND PASSENGER COMFORT

13 *44%* RAILWAYS FROM 1914 TO 1945

13

27
CHILDREN'S PAINTINGS *57%*

28
TAPE AND SLIDE THEATRE *52%*

29

"RAILWAYANA"

29 *64%*

29

29

29

"RAILWAYANA"

29

30

30 RAILWAYS AND SHIPS *69%*

STAIRS

25 TOY RAILWAYS *52%*

25

14 1945 TO THE LAST DAYS OF STEAM

14 *45%*

15 TRAINS FOR EXPORT

30%

29% **16** INDUSTRIAL RAILWAYS

32% **17** MODERN SIGNALLING

WORKS PLATES AND MODELS **26** *24%*

34% **18** DIESEL TRACTION AND ELECTRIFICATION

19 HIGH SPEED TRAINS

58%

24% **20** PROFITABLE FREIGHT

21 DRUM DIGGER *31%*

39% **22** PERMANENT WAY AND LONDON UNDERGROUND TUNNEL

31%

23 TOWARDS TOMORROW

Each case gives a short general account of its subject in bold print; more detailed accounts are given in rather smaller print; and there are also labels by individual exhibits. Amongst the sections which a good many of the visitors recalled looking at were the section on very early beginnings (item 1, looked at by 53 per cent of visitors), the diorama of the opening of the Stockton and Darlington Railway (item 3, 56 per cent), the cases devoted to mid-Victorian railways (item 10, 48 per cent), and the section relating to high speed trains (item 19, 58 per cent). Occasional cases at the side of the balcony overlooking the hall deal with tickets, works-plates and models, and toy railways.

At the end of this section the balcony turns two corners taking it slightly more into the body of the hall. At the wall side by the first corner is an exhibition of children's paintings (item 27) which had attracted the attention of 57 per cent of the visitors we interviewed. Fifty-two per cent had gone into the tape and slide theatre (item 28) which showed slides accompanied by a commentary and music. The remaining two sections 'Railwayana' (item 29) and 'Railways and Ships' (item 30) contained a varied assortment of items including models and old railway company regalia; about two thirds of the balcony's visitors had looked at each.

Tables 4.7 and 4.8 show how the visitors rated each of the 'galleries' they were just leaving for interest and enjoyment. The two tables both show each gallery being given the highest possible rating by over half the sample. The ratings given to the Main Hall were particularly positive. Sixty per cent of the people contacted as they left the Main Hall had spent an hour or more there, as might be expected since it constitutes the main body of the museum. However, many visitors devoted a considerable time to the balcony as well. Table 4.9 indicates that only 14 per cent of the balcony sample had spent less than 15 minutes there, while 43 per cent had taken half an hour or more to look at the exhibits and visit the tape and slide theatre.

Table 4.7   Interest levels in the two National Railway Museum galleries where there were interviews

	Balcony	Main Hall
	%	%
Very interested	53	67
Fairly interested	44	32
Not really interested	3	2
Base	248	263

Source of table: the quota samples interviewed as they left selected galleries.

Table 4.8   How enjoyable visitors found the two National Railway Museum galleries where there were interviews

	Balcony	Main Hall
	%	%
Very enjoyable	56	71
Fairly enjoyable	43	28
Not really enjoyable	1	1
Base	248	263

Source of table: the quota samples interviewed as they left selected galleries.

Table 4.9   Time spent by visitors in the two National Railway Museum galleries where there were interviews

Time spent in gallery	Balcony	Main Hall
	%	%
5 minutes or less	3	0
6–14 minutes	11	0
15–29 minutes	**43**	10
30–59 minutes	33	29
One hour or more	10	**60**
Base	248	263

The percentages in bold indicate the time range containing the median length of time spent in each gallery.

Source of table: the quota samples interviewed as they left selected galleries

d. Measures of interest and enjoyment

Each visitor interviewed on leaving one of the galleries selected for special study was asked two questions, one about their level of interest—*"Did you find* [the gallery in question] *very interesting, fairly interesting, or not really interesting?"* and one about their level of enjoyment— *"Did you find* [the gallery in question] *very enjoyable, fairly enjoyable, or not really enjoyable?"*. Tables 4.1, 4.4 and 4.7 and Tables 4.2, 4.5 and 4.8 showed how these two questions were answered in each of the ten selected galleries. The patterns of replies in the two sets of tables are very similar. This was because most visitors—73 per cent or more in each gallery—gave the same rating in response to both questions. The fact that the ratings given for interest and for enjoyment are so strongly linked means that in the remaining chapters about visitors' reactions to specific galleries we can concentrate on the factors relating to interest, secure in the knowledge that what goes for interest levels is very likely to go for levels of enjoyment as well.

5   Initial interest and reactions to exhibits in selected galleries

a. Initial interest

Chapter 4 discussed visitors' enjoyment of the ten selected galleries and how they rated the galleries for interest. Interest levels will be a recurrent theme of the next few chapters. They will attempt to identify methods of presentation which are successful in enhancing visitors' interest and also to identify which visitors find the galleries particularly interesting and which find them less so. The present chapter is concerned with visitors' interest in the exhibits themselves and in the gallery as a whole.

Of course the interest a visitor feels in a particular gallery is not simply a result of the exhibits he or she finds there or of the way in which they are presented. The visitor's reaction to the gallery will also relate, possibly in complex ways, to the amount of interest he or she initially brings to the gallery. In order to understand the connection between any gallery's main features and the level of interest reported by visitors as they leave it, we need some measure of their initial interest before visiting it on that day.

A convenient indicator of initial interest was provided by the question *"Have you been visiting (the relevant gallery) because you came across it while walking round the museum or because you had already decided to*

visit it?". The first column of Table 5.1 gives the percentage of those visitors who had simply come across each gallery who reported on leaving it that they had found it very interesting.

Table 5.1   Percentage of visitors rating each gallery 'very interesting' by whether they had deliberately decided to visit it

Gallery	Came across the gallery (a)	Decided to visit the gallery (b)	Bases (a)	(b)
	% saying 'very interesting'			
Victoria and Albert Museum				
Art of China and Japan	45%	73%	143	77
British Sculpture	34%	65%	192	34
Tudor Art	39%	71%	184	51
Continental 17th Century Art	45%	70%	169	53
Science Museum				
Aeronautics	50%	74%	105	118
Printing and Paper	26%	61%	149	41
Time Measurement	24%	61%	148	54
Exploration	62%	73%	159	56
National Railway Museum				
Balcony	50%	58%	128	86

The Main Hall at the National Railway Museum is not included since visitors were not asked whether they had decided to visit it.

School party visitors are excluded for the same reason.

Source of table: the quota samples interviewed as they left selected galleries.

Table 5.2   Time spent by visitors in each Victoria and Albert Museum gallery by whether they had deliberately decided to visit it and by interest

Time spent in each gallery	Came across it			Decided to visit		
	Found gallery:					
	Very interesting	Fairly interesting	Not really interesting	Very interesting	Fairly interesting	Not really interesting
	%	%	%	%	%	%
Art of China and Japan						
5 minutes or less	9	39	58	7	10	—
6–14 minutes	34	30	25	14	24	—
15–29 minutes	36	25	17	38	57	—
30 minutes or more	20	6	0	41	10	—
Base	*64*	*67*	*12*	*56*	*21*	*0*
British Sculpture						
5 minutes or less	23	43	56	5	9	—
6–14 minutes	28	30	33	14	27	—
15–29 minutes	28	20	6	23	36	—
30 minutes or more	20	7	6	59	27	—
Base	*64*	*109*	*18*	*22*	*11*	*1*
Tudor Art						
5 minutes or less	18	37	67	11	0	—
6–14 minutes	23	31	22	22	42	—
15–29 minutes	46	19	11	31	42	—
30 minutes or more	12	14	0	36	17	—
Base	*71*	*95*	*18*	*36*	*12*	*3*
Continental 17th Century Art						
5 minutes or less	3	6	14	8	13	—
6–14 minutes	11	16	43	5	13	—
15–29 minutes	34	47	43	16	19	—
30 minutes or more	52	31	0	70	56	—
Base	*76*	*79*	*14*	*37*	*16*	*0*

School party visitors are excluded.

The second column gives equivalent figures for visitors who had decided, either during or before their visit to the museum, to set out and find the gallery in question. Comparing the two columns we see that the greater initial interest of the visitors who made a decision to visit each gallery is reflected in much higher reported levels of interest on leaving.

The Main Hall at the National Railway Museum is omitted from Table 5.1. This is because it was felt inappropriate to ask its visitors whether they had meant to visit what is really less a gallery than the core of the whole museum.

b. Length and thoroughness of gallery visits
The time visitors spend in each gallery is strongly related both to whether or not they deliberately decided to visit the gallery and to how interesting they find it once they have arrived. As would be expected the more interested the visitor the longer he or she spends in the gallery. This holds both for visitors who simply happened across the gallery in question and for visitors who had set out deliberately to visit it. However visitors who have made a decision to visit the gallery tend to show rather more stamina that those who have simply wandered into it. Among the visitors who found each gallery very interesting those who had decided to visit it generally stayed longer than the others. Deliberate visitors who found that the gallery was only fairly interesting were often willing to give it a little more time than casual visitors reporting the same limited amount of interest. The figures are set out in Tables 5.2 and 5.3.

Table 5.3 Time spent by visitors to each gallery at the Science and National Railway Museums by whether they had deliberately decided to visit it and by interest

Time spent in each gallery	Came across it			Decided to visit		
	Found gallery:					
	Very interesting	Fairly interesting	Not really interesting	Very interesting	Fairly interesting	Not really interesting
	%	%	%	%	%	%
Science Museum						
Aeronautics						
5 minutes or less	4	30	—	3	0	—
6–14 minutes	10	23	—	11	21	—
15–29 minutes	41	23	—	30	36	—
30 minutes or more	45	25	—	55	43	—
Base	*51*	*44*	*8*	*87*	*28*	*2*
Printing and Paper						
5 minutes or less	13	25	47	12	40	—
6–14 minutes	38	43	29	20	20	—
15–29 minutes	36	29	18	32	27	—
30 minutes or more	13	3	6	36	14	—
Base	*39*	*93*	*17*	*25*	*15*	*1*
Time Measurement						
5 minutes or less	11	26	67	12	5	—
6–14 minutes	28	40	13	9	29	—
15–29 minutes	44	25	13	36	38	—
30 minutes or more	17	9	7	42	29	—
Base	*36*	*97*	*15*	*33*	*21*	*0*
Exploration						
5 minutes or less	7	7	—	0	0	—
6–14 minutes	12	22	—	10	13	—
15–29 minutes	29	24	—	22	27	—
30 minutes or more	51	46	—	69	60	—
Base	*99*	*54*	*6*	*41*	*15*	*0*
National Railway Museum						
Balcony						
5 minutes or less	2	4	—	0	6	—
6–14 minutes	5	16	—	10	14	—
15–29 minutes	39	49	—	40	46	—
30 minutes or more	55	32	—	50	34	—
Base	*64*	*57*	*7*	*50*	*35*	*1*

	Found gallery:		
	Very interesting	Fairly interesting	Not really interesting
	%	%	%
Main Hall			
5 minutes or less	0	0	—
6–14 minutes	1	0	—
15–29 minutes	8	13	—
30 minutes or more	91	87	—
Base	*175*	*84*	*4*

School party visitors are excluded, except in the case of the Main Hall at the National Railway Museum.

The more time visitors spent in each gallery the more they were able to see. However few, even among those who spent a comparatively long time in each gallery, attempted to take in anything like all the exhibits it contained. Table 5.4 and 5.5 show how many items visitors who had spent half an hour or more in each gallery reported stopping to look at. The 'items' concerned are the numbered sections on our detailed gallery plans. Many of these items of course contain a number of separate exhibits. The plan of the Tudor Art gallery shows 28 different items but, even after spending half an hour or more there, only 15 per cent of its visitors said that they had stopped to look at more than 16 items. Over half had stopped to look at ten items or less. The figures for the other three galleries shown in Table 5.4 are similar.

The last two columns of Table 5.5 suggest that this point should not perhaps be made quite so emphatically in the case of the National Railway Museum. The great majority of visitors interviewed in the Main Hall—234 out of 263—had spent half an hour or more looking round it (most of these in fact spent an hour or more there) and Table 5.5 shows that 51 per cent of them had stopped to look at over 15 different items. The explanation of this comparatively comprehensive coverage is probably the amount of time visitors devote to the Main Hall which really dominates the whole museum. It is less easy to see why 47 per cent of the people who visited the Balcony at the National Railway Museum for half an hour or more should have had time to stop and look at over 15 of the items there. It is possible that these 'items' contained rather less to see than some of the 'items' defined by the plans of the galleries at the two London museums.

Table 5.4 Number of items looked at by visitors spending 30 minutes or more in each Victoria and Albert Museum gallery

Number of items looked at	Art of China and Japan	British Sculpture	Tudor Art	Continental 17th Century Art
	%	%	%	%
1–3	5	5	0	3
4–6	7	16	30	20
7–10	31	43	38	34
11–15	31	16	18	32
16 or more	26	19	15	11
Base	*42*	*37*	*40*	*103*

Source of table: the quota samples interviewed as they left selected galleries.

Turning to Table 5.5 we see that, in the Science Museum too, very few visitors attempt to stop and look at all, or even nearly all, the exhibits in a gallery. In the Aeronautics gallery, although 37 separate items were marked on our map, only 23 per cent of the visitors who had spent half an hour or more there reported looking at more than 15 of the items.

Table 5.5 Number of items looked at by visitors spending 30 minutes or more in each gallery at the Science and National Railway Museums

Number of items looked at	Science Museum				National Railway Museum	
	Aeronautics	Printing and Paper	Time Measurement	Exploration	Balcony	Main Hall
	%	%	%	%	%	%
1–3	6	9	5	4	3	3
4–6	22	18	16	30	8	6
7–10	30	36	35	33	13	18
11–15	19	27	27	26	30	22
16 or more	23	9	16	7	47	51
Base	*108*	*22*	*37*	*125*	*105*	*234*

Source of table: the quota samples interviewed as they left selected galleries.

c. Interest in individual exhibits

Visitors were asked whether they had found any particular exhibit in the gallery they were leaving especially interesting. Tables 5.6 and 5.7 list the items which figured most often in their replies. These correspond fairly closely to the exhibits which attracted most attention generally, in the sense of attracting people to stop and look at them (see figures given on the gallery plans). However there are some interesting exceptions to this rule. The statues of *Raving* and *Melancholy Madness* in the British Sculpture gallery were each looked at by only a few more visitors than the monument to a husband and wife, but almost three times as many people nominated one or both of them as the most interesting item in the gallery. In the Continental Seventeenth Century Art gallery the rather curious miniature wax sculptures scored highest for interest even though there were several other items which more visitors had stopped to view.

In the Aeronautics gallery the *Spitfire,* the exhibit which attracted most attention, was also the exhibit which was most often mentioned as being especially interesting. However the *Hurricane* and the *Vickers Vimy,* the two exhibits which came closest to it in their power to attract attention were named as particularly interesting by fewer people than picked out one or more of the cockpits, or one or more of the piston engine exhibits. It may be that many of the people who found the *Hurricane* and *Vimy* interesting failed to mention them because they found the *Spitfire* more

so, while the cockpits and engines appealed particularly to other sets of visitors.

In the Exploration gallery more than twice as many visitors found the *Apollo 10* spacecraft especially interesting as picked out the lunar lander model although the same proportion of visitors stopped to look at each. The thermal camera display received the second highest number of mentions for being especially interesting despite only coming fifth in terms of the number stopping to look at it. In the Time Measurement gallery the watches attracted surprisingly few citations for being particularly interesting. This may be because of the form in which the question was asked. Visitors were asked if they could name one object which they had found especially interesting. People who found the watches as a whole interesting might not have felt able to pick out one particular watch.

Tables 5.6 and 5.7 also make it possible to compare the tastes of the visitors whose interest in each gallery was serious enough for them to make a definite decision to visit it with those of the visitors who merely happened on the gallery. It might be expected that the exhibits which appealed most to visitors with a definite interest in a gallery would differ from those which particularly struck casual visitors. However Tables 5.6 and 5.7 do not lend this conjecture much support. Although there are some apparent differences in particular galleries between the answers given by people

Table 5.6 The items found especially interesting by visitors to each Victoria and Albert Museum gallery by whether they had deliberately decided to visit the gallery

Item(s) selected as most interesting	Came across the gallery	Decided to visit the gallery	All visitors
	%	%	%
Art of China and Japan			
Item 20, Ch'ien-Lung throne	22	2	16
Item 14, Throne and screen	16	11	15
Item 8, Pottery horses and figures	16	13	14
Item(s) selected, but not one of the three most often named	47	74	56
Base	*108*	*61*	*179*
British Sculpture			
Items 24 and 25, *Raving* and *Melancholy Madness*	33	29	31
Item 28, Monument to husband and wife	12	7	11
Item 7, *St Michael and Satan*	7	11	8
Item(s) selected, but not one of the three most often named	48	54	49
Base	*139*	*28*	*179*
Tudor Art			
Item 13, Great Bed of Ware	59	43	55
Item 19, Jacket and portrait	8	15	11
Item 10, Panelled room	8	10	9
Item(s) selected, but not one of the three most often named	25	33	26
Base	*156*	*40*	*211*
Continental 17th Century Art			
Item 16, Wax sculpture	18	7	16
Item 26, Small scale furnishings	16	15	15
Item 22, Sculpture	7	29	13
Item(s) selected, but not one of the three most often names	59	49	57
Base	*118*	*41*	*168*

Percentages may add to slightly over 100% as some visitors selected more than one of the three most often named items.

Visitors who felt unable to single out any item as particularly interesting are excluded from this table. School party visitors are excluded from the first two columns.

Source of table: the quota samples interviewed as they left selected galleries.

Table 5.7 The items found especially interesting by visitors to each gallery at the Science and National Railway Museums by whether they had deliberately decided to visit the gallery

Item(s) selected as most interesting	Came across the gallery	Decided to visit the gallery	All visitors
	%	%	%
Science Museum			
Aeronautics			
Item 36: *Spitfire*	14	15	15
Item 10: Cockpits	12	13	11
Item 5: Piston engines	7	12	10
Item(s) selected but not one of the three most often named	68	60	64
Base	*59*	*86*	*168*
Printing and Paper			
Item 17: Printing machines	25	28	29
Item 8: Mechanical typesetting	30	9	24
Item 2: Hand presses	9	13	9
Item(s) selected but not one of the three most often names	35	50	38
Base	*99*	*32*	*154*
Time Measurement			
Item 12: Rolling ball clock	39	17	34
Item 9: Wells Cathedral clock mechanism	14	29	17
Item 19: Watches	5	6	5
Item(s) selected but not one of the three most often named	42	48	44
Base	*118*	*48*	*186*
Exploration			
Item 10: *Apollo 10* spacecraft	33	51	38
Item 17: Thermal camera display	18	16	17
Item 9: Lunar lander model	15	8	14
Item(s) selected but not one of the three most often named	34	24	31
Base	*135*	*49*	*208*
National Railway Museum			
Balcony			
Item 30, Railways and ships	18	11	15
Item 28, Tape and slide theatre	12	10	11
Item 29, Railwayana	8	8	9
Item(s) selected but not one of three most often mentioned	62	71	64
Base	*90*	*63*	*181*
Main Hall			
Item 18, Royal coaches			25
Item 6, *Mallard*			16
Item 13, Sectioned locomotive			10
Item(s) selected but not one of three most often mentioned			48
Base			*219*

Percentages may add to slightly over 100% as some visitors selected more than one of the three most often named items.

Visitors who felt unable to single out any item as particularly interesting are excluded from this table. School parties are excluded from the first two columns.

Source of table: the quota samples interviewed as they left selected galleries.

who drifted into them and those given by people who set out to find the gallery, it is hard to discern the kind of consistent pattern that would suggest that we were dealing with much more than random variation.

The question of which exhibits were found most interesting leads on to a more general question, namely what is it that people actually notice about the objects they look at during their visits. In order to investigate this, interviewers in each gallery were given six show cards each consisting of a fairly large photograph of an exhibit which, it was expected, would have attracted the attention of a considerable proportion of the gallery's visitors. Visitors were shown these cards and asked to indicate which of the six exhibits, if any, they had looked at most closely during their visit. They were then asked the following three questions about the exhibit:

"I would like you to tell me about the object(s) in this picture. To start with can you tell me what it is (they are)?"

"What do you think are the main points of interest about it (them)?"

"Is there anything else you can tell me about it (them)?"

The points they made about the exhibits were condensed into the 15 categories shown in Table 5.8.

The results can be set out separately for individual exhibits as is done by Figure 5.1. However the general pattern emerges most clearly when the results for all six exhibits in each gallery are combined. Table 5.8 presents the results in this form. Table 5.9 gives the basis for these figures by listing the six relevant exhibits in each gallery and giving the number of visitors who selected and commented on each exhibit.

Figure 5.1 Comment profiles of 10 exhibits - proportion of informants making various kinds of comment

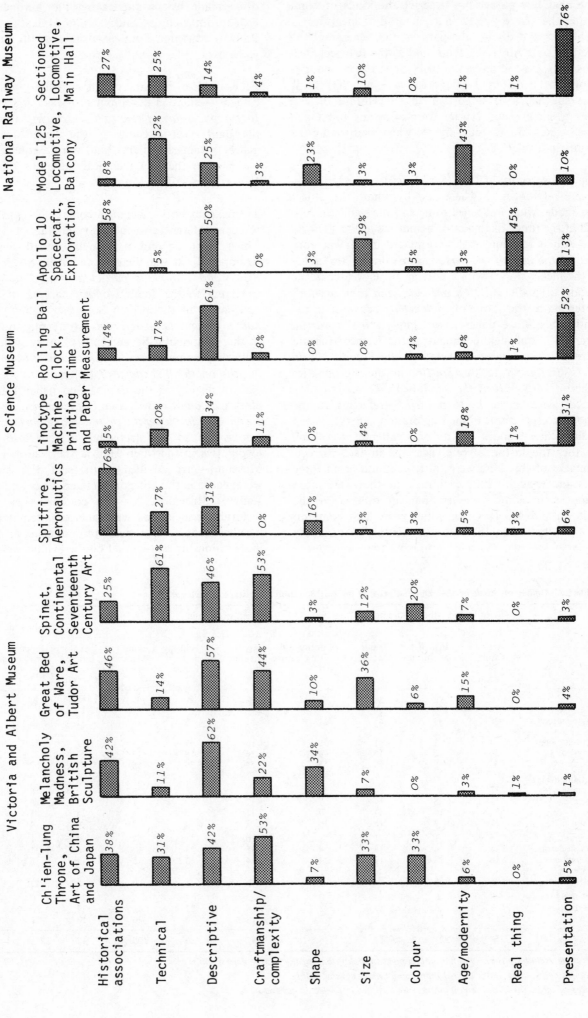

The next few paragraphs describe the kinds of comments that were placed in each of the categories in Table 5.8 and discuss the patterns that emerge at the Victoria and Albert Museum and at the two technical museums. As the discussion will show, the boundaries between the different categories were not very clear cut, so that it is important to concentrate on the general picture and refrain from attaching too much significance to individual figures when interpreting the results in Table 5.8.

The first category in Table 5.8 historical associations—covers references to where and by whom the object was made, what happened to it, and the significance it had at the time. Examples of comments at the Victoria and Albert Museum that were placed in this category are: *"I thought of the person who wore it"* (referring to the jacket shown with a portrait of its wearer in the Tudor Art gallery); *"It's on loan from the hospital"* (referring to the statue of *Melancholy Madness* in the British Sculpture gallery); or simply *"It's historical interest"*. Equivalent comments at the Science Museum were: *"It's an aeroplane used in the Second World War"* (referring to the *Spitfire* in the Aeronautics gallery); *"It's part of the Apollo capsule—part of the rocket which came back to earth"* (referring to the *Apollo 10* spacecraft in the Exploration gallery). When coding these replies we did not attempt to check whether the historical reference was in fact correct. Remarks of this kind were fairly common in all three museums though within each museum they were more common in some galleries than in others. In the Victoria and Albert Museum they were least common in the Continental Seventeenth Century Art gallery while in the two technical museums they were more

often made by people leaving the Aeronautics and Exploration galleries, and the Main Hall at the National Railway Museum, than by visitors to the other three galleries.

An example of the kind of answer placed in the second category—stage in development—was *"It's one of the first applications of electricity in clocks"* (referring to the Bain Electric Clock in the Time Measurement gallery). Remarks of this kind were a good deal more common in the two technical museums than in the Victoria and Albert Museum.

The next category—technical comments—was intended to cover references to factual points about the exhibit which went beyond a straightforward description. Examples from the Victoria and Albert Museum are: *"Boxwood I think is particularly hard to work in"* (referring to the small wooden carving of the Last Judgement in the British Sculpture gallery) and *"It had an equally tempered scale"* (referring to the spinet in the Continental Seventeenth Century gallery). The following comments—*"It picks up different wavelengths on the TV camera"* (referring to the thermal camera display in the Exploration gallery) and *"The electric locomotive was used in the tunnel, because the steam couldn't operate in the tunnel because of the smoke and so on"* (referring to the *Bo-Bo* electric locomotive in the Main Hall at the National Railway Museum)—give an idea of the sort of answers that were placed in this category in the Science and National Railway Museums. As with comments on exhibits' historical associations we made no attempt to check whether technical comments or comments on the stage in development represented by the exhibit were correct

Table 5.8 Comments made on the exhibit selected at the card question by visitors to each gallery

	Victoria and Albert Museum				Science Museum				National Railway Museum	
	Art of China and Japan	British Sculpture	Tudor Art	Continental 17th Century Art	Aero-nautics	Printing and Paper	Time Measure-ment	Explor-ation	Balcony	Main Hall
	%	%	%	%	%	%	%	%	%	%
Historical associations	36	34	41	21	35	8	18	31	15	36
Stage in development	4	1	1	2	16	7	4	12	14	6
Technical	30	12	15	46	16	18	15	26	27	24
Descriptive	44	48	60	55	54	39	54	47	32	24
Craftsmanship/ complexity	38	29	44	59	2	8	8	0	6	9
Shape	13	28	14	9	7	2	1	3	8	16
Size	24	14	34	9	11	5	8	24	1	9
Colour	39	4	13	10	1	0	2	4	4	7
Preservation	6	1	11	1	0	1	7	2	1	10
Age/modernity	8	4	15	4	16	23	17	8	32	9
Evaluative	16	28	17	24	4	2	15	3	3	4
Personal association	6	6	7	4	6	5	6	5	13	4
Real thing	0	0	1	0	2	1	1	20	0	1
Presentation	3	4	13	2	8	22	41	27	19	38
Other	11	11	8	12	13	20	20	10	11	16
Base	218	225	242	226	247	180	196	237	225	254

This table excludes visitors who had not stopped to look at any of the picture card exhibits.

Columns add to over 100% as informants could make several points.

Source of table: the quota samples interviewed as they left selected galleries.

Table 5.9 List of the show card exhibits by gallery, showing positions on gallery plans and the number of visitors selecting each exhibit.

Exhibit	No. of selections	Exhibit	No. of selections	Exhibit	No. of selections	Exhibit	No. of selections
Victoria and Albert Museum							
Art of China and Japan		**British Sculpture**		**Tudor Art**		**Continental 17th century Art**	
Screen (part of item 25)	15	Monument to husband and wife (item 28)	75	Tapestry (part of item 4)	3	Bust (part of item 22)	12
Ch'ien-Lung throne (item 20)	88	Melancholy Madness (item 25)	89	Great Bed of Ware (item 13)	155	Cabinet (part of item 12)	48
Pottery horse (part of item 8)	62	Handel (item 26)	18	Bust of Henry VII (item 2)	2	Time and Death (part of item 16)	53
Vase (near item 9)	11	Last Judgement (part of item 19)	20	Jacket (part of item 19)	59	Silver clock (item 3)	38
Buddha (part of item 12)	10	Sketch model for monument (item 4)	9	Virginals (item 11)	22	Virgin of Sorrows (item 6)	16
Robe (part of item 16)	32	Valour and Cowardice (item 1)	14	Tapestry fragment (part of item 24)	1	Spinet (part of item 26)	59
Science Museum							
Aeronautics		**Printing and Paper**		**Time Measurement**		**Exploration**	
Wright flyer copy (item 16)	37	Stanhope iron press (part of item 2)	9	Foliot clock mechanism (part of item 8)	14	X-ray and UV satelites (item 22)	1
Whittle engine (part of item 12)	26	Lumitype photosetter (part of item 14)	10	Atomic clock (item 32)	9	Diving chamber reconstruction (item 13)	30
Spitfire (item 36)	62	Linotype machine (part of item 8)	110	Bain electric clock (part of item 30)	18	Thermal camera display (item 17)	41
Piccard's Gondola (part of item 7)	26	Model rag boiler (part of item 11)	17	Model of Chinese water clock (part of item 4)	11	Douglas fir (item 15)	27
Vickers Vimy (item 28)	39	Model Koenig and Bauer press (part of item 17)	21	Wells Cathedral clock mechanism (item 9)	58	Apollo 10 spacecraft (item 10)	109
Gloster Meteor cockpit (part of item 10)	57	Early typewriter (part of item 22)	13	Rolling ball clock (item 12)	86	EMI scanner (item 8)	29
National Railway Museum							
Balcony		**Main Hall**					
Model of medieval mine workings (part of item 1)	52	Sectioned locomotive (item 13)	100				
Railway tickets (part of item 8)	30	Queen Victoria's coach (part of item 18)	38				
Model Garratt locomotive (part of item 15)	15	Bo-Bo electric locomotive (part of item 27)	9				
Model 125 locomotive (part of item 19)	60	Piston value models (part of item 28)	15				
Model drum digger (item 21)	32	Mallard locomotive (item 6)	78				
Group of models (part of item 29)	36	Signalling and points display (part of item 33)	14				

Source of table: the quota samples interviewed as they left selected galleries.

before placing them in the appropriate categories. Technical comments, defined in this way, were considerably more common than comments referring to stages of development. They were as common in the Victoria and Albert Museum as in the other two museums, though naturally the kind of technical point being made would have been different in the different museums.

The fourth category—descriptive comments—covered more answers than any other in seven of the ten galleries and was among the most common in the other three. It included factual points which were not felt to call for any technical understanding, though as some of the following examples show the line between descriptive and technical comments could in practice be

rather thin. Comments placed in this category at the Victoria and Albert Museum included *"It's carved"*, *"It's some sort of pottery"*, *"I wondered how they got it into a room"* (referring to the Great Bed of Ware in the Tudor Art gallery) and *"The body is half eaten away so you can see the intestines, a skeleton is standing by"* (referring to the wax tableau of Time and Death in the Continental Seventeenth Century gallery). Comments that were classified as descriptive at the two technical museums included *"It goes under the water"* (referring incorrectly to Piccard's balloon gondola in the Aeronautics gallery), *"It's a mechanical typesetting equipment"* (referring to a linotype machine in the Printing and Paper gallery) and *"The style and design plus increased efficiency"* which the speaker considered notable features of the 125 high speed locomotive, a model of which was one of the show card items for the Balcony at the National Railway Museum.

It is the next four categories which show most clearly the difference between the type of interest which visitors bring to the Victoria and Albert Museum and the way they look at exhibits in the two technical museums. Visitors at the Victoria and Albert Museum show an interest in the appearance of the exhibits which goes far beyond bald description. Comments on the craftsmanship involved in making the exhibits include references to *"the expertise in making it"* and the *"delicacy of the carving"* as well as comments such as *"the detail . . . . you can see all the muscles, the ribs"*. Comments on the shapes of exhibits range from that of the visitor who explained the appeal of an exhibit as being *"probably the neatness of the shape"* to that of a visitor who remarked on an exhibit's *"symmetry and artistic balance, arrangement of lines"*.

Comments on size included such remarks as *"A very large bed"* (referring to the Great Bed of Ware in the Tudor Art gallery); *"The size principally . . . . . . sort of large"* and *"All the figures are so small"*. The comments that were placed in the 'colour' category are ones that showed a definite interest in the exhibit's colour, for instance *"The colours they used on the cream background. They're not drab"* or *"The colour is fantastic"* but not a simple statement of fact such as *"It is red"* which would have been placed in the category of 'descriptive' comments.

Turning to the comments made in the galleries at the two technical museums we see that their visitors make far fewer references to the shape and size of the exhibits, or to the effect of their colouring or the craftsmanship that went into them, than do the visitors to the Victoria and Albert Museum.

Only two other categories of answer call for comment here. One is the 'evaluative' category, which covers references to the beauty of exhibits or statements that the visitor personally likes or dislikes them. Answers of this type are also far more common at the Victoria and Albert Museum than at the Science and National Railway Museums.

Finally we have the result that visitors to the two technical museums make more comments on the way in which the exhibits are presented than do visitors to the Victoria and Albert Museum. Many of these answers relate to special effects such as exhibits which are actually in motion.

It is now time to consider the comments made on individual exhibits. Figure 5.1 gives 'comment profiles' of the exhibits that were most often selected at the show card question. The comments are divided into the same categories as in Table 5.8 except that some of the less frequent categories have been omitted. These comment profiles should give some idea of the reasons for the appeal of the individual exhibits.

The four exhibits in the Victoria and Albert Museum have broadly similar comment profiles. All attracted substantial numbers of general descriptive comments, remarks on the historical associations of the exhibit in question, and comments on its complexity or on the craftsmanship involved in making it. Each exhibit also attracted a number of comments under one or more of the headings 'shape', 'size' or 'colour'. Thus all four exhibits seem to have appealed for both historical and aesthetic reasons. Technical points were made by 61 per cent of the visitors who commented on the spinet and by 31 per cent of the people who described the Ch'ien-lung throne.

The profiles shown by the exhibits at the two technical museums are more varied. To judge from visitors' remarks, the Spitfire's appeal is overwhelmingly due to its role in the Second World War. More recent historical associations are important in the case of the *Apollo 10* spacecraft, which was used for a manned flight round the moon. Forty-five per cent of the people who commented on the spacecraft expressed some awe at the fact that it was this particular object which had done something so remarkable, that this was the real thing or as one visitor said *"It's the actual piece of equipment"*. It was the only exhibit in any of the ten galleries to evoke this reaction. Perhaps it derived its appeal because many people, having recently felt intensely but only vicariously involved in the enterprise of placing men on the moon, enjoyed seeing their somewhat remote experience embodied in a tangible object. As one visitor said, *"I . . . . . . collect photos of space travel and wanted to see the real thing"*.

The rolling-ball clock is so called because the timing device conceived by its designer Congreve consists of a ball rolling down a zig-zag groove on an inclined plane. The large number of descriptive comments confirm the interest aroused by its strange design. Over half the visitors commenting on it remarked on the way it was displayed—the most notable feature of which was that, like several other exhibits in the Time Measurement gallery, it was actually in operation. The same was often true of the linotype machine in the Printing and Paper gallery.

The interest shown in the sectioned locomotive in the Main Hall of the National Railway Museum was also due to the way it was presented—as the remarks of 76 per cent of the visitors who described it made clear. One side of the locomotive had been cut away to give visitors a view of the contents of the boiler. The model 125 locomotive in the Balcony at the National Railway Museum illustrates an extensive explanatory text. This may account for the prominence of technical remarks in its comment profile.

6 Presentation

a. The importance of attractive presentation

Visitors to each gallery were asked whether, when they first walked into it, they had found its general appearance 'very attractive', 'fairly attractive' or 'not really attractive.' Table 6.1 sets out their answers. In the Victoria and Albert Museum the gallery which received the highest rating of the four for attractiveness was the Continental Seventeenth Century Art gallery. In the Science Museum the Exploration and Aeronautics galleries were more often thought to be very attractive than were the other two galleries selected for special study. The Main Hall at the National Railway Museum was considered very attractive by a higher proportion of its visitors than the Balcony overlooking it. The gallery which seems, of the ten where interviewing was conducted, to have struck its visitors as least attractive was the Time Measurement gallery. Only 18 per cent of its visitors found it very attractive, compared to 26 per cent who confessed to finding it not really attractive. This pattern of attractiveness ratings is very similar to the pattern of interest level ratings shown earlier in Tables 4.1, 4.4 and 4.7—suggesting that there may be a close connection between the perceived attractiveness of a gallery and the amount of interest it arouses. As we shall see, the connection seems to work both ways.

Table 6.2 shows that visitors who have a prior interest in the subject of a gallery are likely to find it more attractive than those who have not. The Art of China and Japan gallery can serve as an example, where 47 per cent of the people who had deliberately set out to find the gallery thought it very attractive, compared to only 34 per cent of the people who had merely come across it.

Table 6.3 relates visitors' opinions of each gallery's appearance to the interest they felt once they were there. If we look at the first three columns of the table,

Table 6.2 Percentage of visitors rating each gallery 'very attractive' by whether they had deliberately decided to visit the gallery

Gallery	Came across the gallery (a)	Decided to visit the gallery (b)	Bases (a)	(b)
Victoria and Albert Museum	*% saying 'very attractive'*			
Art of China and Japan	34%	47%	143	77
British Sculpture	32%	44%	192	34
Tudor Art	26%	38%	184	51
Continental 17th Century Art	54%	57%	169	53
Science Museum				
Aeronautics	46%	57%	105	118
Printing and Paper	34%	63%	149	41
Time Measurement	15%	23%	148	54
Exploration	52%	59%	159	56
National Railway Museum				
Balcony	40%	59%	128	86

The Main Hall at the National Railway Museum is not included since visitors were not asked whether they had decided to visit it.

School party visitors are excluded for the same reason.

Source of table: the quota samples interviewed as they left selected galleries.

which exclude the visitors who had purposely decided to visit each gallery, and compare the interest ratings given by those who found the gallery very attractive with the interest reported by those who found it fairly or not really attractive a clear picture emerges. In every case except one the visitors who most liked the appearance of the gallery rated it more highly for interest than did other visitors.

This conclusion suggests that attractive presentation can greatly enhance the interest visitors feel. However it could imply less than it seems to. A gallery might be judged attractive not because of the way it was set out but because of the appeal of the exhibits themselves. Table 6.4, which is based on the replies of casual

Table 6.1 Ratings of the appearance of each gallery

Appearance of gallery	Victoria and Albert Museum				Science Museum				National Railway Museum	
	Art of China and Japan	British Scultpure	Tudor Art	Continental 17th Century Art	Aeronautics	Printing and Paper	Time Measurement	Exploration	Balcony	Main Hall
	%	%	%	%	%	%	%	%	%	%
Very attractive	38	35	30	54	51	37	18	55	47	63
Fairly attractive	46	46	50	38	39	51	55	36	44	31
Not really attractive	16	19	20	7	11	11	26	9	9	6
Base	232	240	251	236	255	222	227	243	248	263

Source of table: the quota samples interviewed as they left selected galleries.

Table 6.3 Percentage of visitors rating each gallery 'very interesting' by whether they had deliberately decided to visit it and how they rated its appearance

Gallery	Came across the gallery			Decided to visit the gallery		
	Thought gallery:					
	Very attractive	Fairly attractive	Not really attractive	Very attractive	Fairly attractive	Not really attractive
Victoria and Albert Museum	% saying 'very interesting'					
Art of China and Japan	64%	38%	29%	78%	72%	60%
Base	*47*	*72*	*21*	*36*	*25*	*15*
British Sculpture	43%	34%	21%	53%	77%	—
Base	*61*	*89*	*39*	*15*	*13*	*6*
Tudor Art	63%	35%	21%	79%	72%	—
Base	*48*	*92*	*42*	*19*	*25*	*6*
Continental 17th Century Art	50%	38%	40%	73%	68%	—
Base	*90*	*68*	*10*	*30*	*19*	*4*
Science Museum						
Aeronautics	63%	39%	—	77%	74%	58%
Base	*48*	*46*	*10**	*67*	*38*	*12*
Printing and Paper	27%	28%	25%	73%	46%	—
Base	*49*	*80*	*16*	*26*	*13*	*2*
Time Measurement	57%	23%	7%	83%	58%	—
Base	*21*	*79*	*44*	*12*	*33*	*8*
Exploration	67%	56%	58%	76%	75%	—
Base	*83*	*64*	*12*	*33*	*16*	*7*
National Railway Museum						
Balcony	63%	42%	38%	69%	43%	—
Base	*51*	*64*	*13*	*51*	*30*	*5*
	Thought Main Hall:					
	Very attractive	Fairly attractive	Not really attractive			
Main Hall	76%	52%	44%			
Base	*165*	*82*	*16*			

* One of these 10 visitors was a 'no answer'.
School party visitors are excluded except in the case of the Main Hall; see footnote to Table 6.2.
Source of table: the quota samples interviewed as they left selected galleries.

visitors, examines this possibility. The visitors who thought each gallery very or fairly attractive are divided into two groups according to whether they gave their interest in the gallery's contents as a reason* for judging it attractive. The table shows what proportion of each group found the galleries very interesting. The interest levels reported by those who did not find the galleries really attractive are also given for comparison.

The figures in Table 6.4 suggest that the appeal of the exhibits themselves possibly lies behind some of the additional interest reported by those visitors who thought that the Art of China and Japan and British Sculpture galleries at the Victoria and Albert Museum and the Balcony at the National Railway Museum were attractive. However in the remaining galleries the interest levels reported by visitors who attributed the galleries' attractiveness purely to presentation were as high as the interest levels of those who thought that the contents played a part. In these galleries then, the increased interest does appear to result from attractive presentation, confirming the importance of the way a gallery is set out.

The attractiveness of a gallery may be more important for some visitors than for others. Returning to Table 6.3 we can compare the importance of attractive presen-

tation for deliberate visitors with its importance for casual visitors. The difference between the interest levels reported by visitors who found the gallery very attractive and those reported by visitors who only found the gallery fairly attractive is usually far more noticeable in the case of people who simply happened on the gallery. This is true of the Victoria and Albert Museum galleries and of the Aeronautics and Exploration galleries at the Science Museum. In the case of the Time Measurement gallery and of the Balcony at the National Railway Museum the difference in interest levels is approximately the same for both sets of visitors. Only the Printing and Paper gallery reverses the pattern. There the perceived attractiveness of the gallery was related to the interest levels of those sample members who were deliberate visitors but not to the interest felt by people who simply came across the gallery. The general pattern of these results suggests that attractive presentation matters less for visitors who start with a strong interest in the subject of a gallery than it does for people whose initial attitude is rather lukewarm.

b. Reactions to the lay-out of individual galleries
i. Victoria and Albert Museum
Table 6.5 sets out the reasons which visitors gave for finding each gallery attractive. The contents themselves, whose contribution to galleries' attractiveness was discussed in the previous section, are often mentioned of course, but our concern here is with the contribution made by the way each gallery is set out. The figures in

---

* Reasons given for finding particular galleries attractive are discussed fully in Section b of this chapter.

Table 6.4 Percentage of visitors rating each gallery 'very interesting' by how they rated its appearance and whether they said the gallery's contents helped to make it attractive—visitors who said they came across the gallery

Gallery	Thought gallery:			Bases		
	Attractive		Not really attractive	(a)	(b)	(c)
	Contents mentioned (a)	Contents not mentioned (b)	(c)			
	% saying 'very interesting'					
Victoria and Albert Museum						
Art of China and Japan	62%	43%	29%	29	90	21
British Sculpture	59%	33%	21%	29	121	39
Tudor Art	45%	44%	21%	47	93	42
Continental 17th Century Art	43%	45%	40%	30	128	10
Science Museum						
Aeronautics	50%	51%	—	18	76	10*
Printing and Paper	27%	27%	25%	26	103	16
Time Measurement	30%	30%	7%	23	77	44
Exploration	63%	62%	58%	30	117	12
National Railway Museum						
Balcony	71%	49%	38%	14	101	13
Main Hall	70%	68%	44%	27	220	16

One of these 10 visitors was a 'no answer'.

The figures for the Main Hall at the National Railway Museum include all sample members since visitors were not asked whether they had decided to visit it.

Source of table: the quota samples interviewed as they left selected galleries.

Table 6.5 Reasons for finding each Victoria and Albert Museum gallery attractive

Reasons for finding the gallery attractive	Art of China and Japan	British Sculpture	Tudor Art	Continental 17th Century Art
	%	%	%	%
Lighting	28	27	12	26
Darkness	0	0	14	6
Colour	10	10	10	15
Spaciousness, lack of clutter	42	41	10	27
Realism, room settings	1	2	23	10
(Chrono)logical sequence	2	0	0	3
Miscellaneous aspects of lay-out	27	19	26	39
The building itself	12	22	3	3
Variety, so much to see	4	11	4	6
Tidy, bright clean	14	15	4	9
Eye-catching	10	12	5	2
Sense of excitement	2	3	4	2
Restfulness	5	3	1	9
Contents	26	23	38	20
Information provided	5	3	2	4
Other	7	6	4	13
Base*	192	191	198	218

Visitors who found the gallery very or fairly attractive.

The percentages add to over 100% as people could give more than one answer.

Source of table: the quote samples interviewed as they left selected galleries.

Table 6.5 are important in two ways. They indicate features of lay-out which particularly appeal to visitors, and they help to show the ways in which the lay-out of each gallery differs from that of the others in the eyes of the visitors.

Spaciousness, or lack of clutter, was mentioned as an attractive feature by just over 40 per cent of the visitors who liked the appearance of the Art of China and Japan and British Sculpture galleries, and by 27 per cent in the Continental Seventeenth Century Art gallery. Answers placed in this category included both general statements that there was plenty of space and the comments of visitors who felt that the way the exhibits were set out made it easy to look at exhibits from different sides or simply made it easy to walk round the gallery. In the case of the Art of China and Japan gallery, which was in fact rather full, it is these latter meanings which are relevant.

The lighting in the Art of China and Japan, British Sculpture, and Continental Seventeenth Century Art galleries was commended by just over a quarter of each gallery's visitors. Few visitors said that the Tudor Art gallery was either spacious or well lit though 14 per cent of the visitors who liked the appearance of the gallery felt that the darkness helped to make it attractive. Perhaps they felt that it complemented the dark wood of the period room settings which were mentioned favourably by 23 per cent.

Twenty-two per cent of the visitors who were pleased by the appearance of the British Sculpture gallery attributed some of its attraction to the structure of the building itself as did 12 per cent of the visitors to the Art of China and Japan gallery. These two galleries were also more often thought to be eye-catching and generally tidy, bright or clean than were the other two. Ten per cent or so of the visitors who liked the appearance of each gallery mentioned the colour either of the background or of the exhibits themselves.

We have left till last the set of comments that have been categorised as referring to 'miscellaneous aspects of lay-out'. This category covers people who said that the gallery was *"well laid out"* without saying why they thought so, the comments of people who said that things in the gallery were *"easy to see"* but did not explain further, and a whole host of specific comments on the way exhibits were positioned. Comments like these were particularly common in the Continental Seventeenth Century Art gallery where they were made by 39 per cent of the visitors who thought the gallery attractive.

So far the positive factors which have stood out are, generally, good lighting and lack of clutter and, in particular galleries, realistic room settings and the structure of the building itself. Strangely enough, none of these factors seem from the evidence of Table 6.5 to distinguish particularly the gallery whose appearance was most often thought very attractive—namely the Continental Seventeenth Century gallery. To discover the secret of its appeal we must turn to Table 6.6 which sets out the comments of the minority of visitors who thought each gallery 'not really attractive'.

In each gallery except the Continental Seventeenth Century Art gallery there were substantial numbers of complaints that the atmosphere of the gallery was impersonal, cold or dreary. Equal numbers remarked on what they felt were the dull colours used in the galleries' decor or simply remarked that they did not contain enough colour. The rather sumptuous design of the Continental Seventeenth Century Art gallery seems to have succeeded precisely because it did not leave people feeling depressed by a chill or colourless atmosphere.

None of the four galleries was thought cramped by many of its visitors, not even the Tudor Art gallery which attracted least positive praise for spaciousness. However, three quarters of the visitors who found the Tudor Art gallery unattractive complained of the lack of light. This is a far higher proportion than that complaining of any single factor in any of the other galleries and suggests that the lack of lighting was seriously limiting the enjoyment of the gallery's visitors.

It is no surprise that 70 per cent of the 120 people who suggested improvements in the presentation of the Tudor Art gallery asked for better lighting (Table 6.7). Improved lighting was a fairly common request in the other three galleries as well. The fact that comparatively

Table 6.6 Reasons for finding each Victoria and Albert Museum gallery unattractive

Reasons for finding the gallery unattractive	Art of China and Japan	British Sculpture	Tudor Art	Continental 17th Century Art
	%	%	%	%
Poor lighting	14	9	76	25
(Lack of) colour	39	51	26	6
Cluttered, cramped	3	11	2	19
Lacks sequence, no planned route	6	13	0	19
Too much to see	0	16	2	19
Glass cases	19	13	4	0
Miscellaneous aspects of lay-out	17	29	14	6
Impersonal atmosphere, dingy, dull	31	31	40	6
Not eye-catching	17	7	4	19
Content	8	7	16	0
Other	42	40	14	44
Base*	36	46	50	17

* *Visitors who found the gallery not really attractive.*
The percentages add to over 100% as people could give more than one answer.
Source of table: the quote samples interviewed as they left selected galleries.

Table 6.7 Improvements suggested to the presentation of each Victoria and Albert Museum gallery

Improvements suggested	Art of China and Japan	British Sculpture	Tudor Art	Continental 17th Century Art
	%	%	%	%
Lighting	27	20	70	30
Colour	9	25	14	3
More space	5	11	2	11
More realism, room settings	6	1	10	11
Design of cases	17	11	2	5
(Chrono)logical sequence	1	0	2	3
Miscellaneous improvements to lay-out	27	44	10	18
Fewer exhibits	1	9	3	8
Different exhibits	4	1	5	3
Information provided	22	8	18	30
Help finding way round	5	7	3	8
Other	27	26	10	21
Base*	81	96	120	68

* *Visitors who thought the presentation could be improved.*
The percentages add to over 100% as people could give more than one answer.
Source of table: the quota samples interviewed as they left selected galleries.

few people suggested making any of the galleries less cluttered is not inconsistent with the importance given to spaciousness as a positive factor; rather, it reflects the fact that most visitors found the four galleries sufficiently spacious already. In general visitors' suggestions for improvements merely serve to confirm their earlier comments on the galleries' appearance.

ii. The Science and National Railway Museums

What might be called 'special effects' seem to have played an important role in making galleries at the Science Museum attractive to their visitors. Looking at Table 6.8 we see that a third of the people who thought the Aeronautics gallery attractive mentioned the aircraft suspended from the ceiling. This multitude of seemingly airborne exhibits may also have been in the minds of the visitors who said they found the gallery attractive because it contained so much to see.

The fact that some of the exhibits were actually working had appealed to 29 per cent at the Printing and Paper gallery. In the Exploration gallery the device of setting brightly coloured and strongly lit exhibits in dark surroundings had obviously appealed to many visitors. Almost a quarter of the visitors who liked the gallery's appearance reported that they had found it eye-catching and ten per cent referred to the sense of excitement they felt. Both the lighting itself and the enveloping darkness received substantial numbers of favourable comments.

A number of visitors praised each of these three galleries for their 'realism'. In the Aeronautics and Printing and Paper galleries this sense of realism is probably due to the special effects. It is less easy to see why the Exploration gallery should be thought realistic. It may

result from the very prominent full-scale model of a landing craft standing on the moon's surface (item 9 on the detailed plan). It is also possible that the darkness itself gives some visitors the sensation of being out in space.

Colour and light contributed greatly to the appeal of the Printing and Paper gallery. Though only 14 per cent of the visitors who liked the gallery's appearance referred specifically to the colour scheme, 19 per cent praised the gallery's lighting and another 19 per cent commended it for being tidy, bright and clean. Ten per cent described the gallery's appearance as eye-catching.

Forty-eight per cent described the Main Hall at the National Railway Museum as tidy, bright or clean. However, important though this obviously is, it is reasonable to suspect that it is not really the main source of the hall's visual appeal. Far fewer visitors bothered to refer to the equally marked cleanliness of the Balcony. The most plausible explanation of the praise heaped on the Main Hall for its cleanliness is that many visitors had not expected the contents of a converted engine shed to be tidy, bright or clean at all. Their answers express a sense of surprise.

Both the Balcony and the Main Hall at the National Railway Museum were much praised for their spaciousness and lack of clutter. As we noted when discussing the results for the Victoria and Albert Museum this category includes both references to the amount of space in the gallery and statements that the way the objects in the gallery were set out gave visitors enough room to see the exhibits properly. It is this second meaning which is important in the case of the Balcony

Table 6.8 Reasons for finding each gallery attractive at the Science and National Railway Museums

Reason for finding the gallery attractive	Science Museum				National Railway Museum	
	Aeronautics	Printing and Paper	Time Measurement	Exploration	Balcony	Main Hall
	%	%	%	%	%	%
Lighting	5	19	7	25	19	13
Darkness	0	0	0	16	0	0
Colour	6	14	0	18	5	13
Spaciousness, lack of clutter	13	14	18	9	31	34
Realism	14	11	1	12	1	17
Turntable, hanging aircraft	35	—	—	—	—	14
Different levels, catwalk, balcony	10	2	1	2	11	0
(Chrono)logical sequence	4	6	11	6	14	2
Miscellaneous aspects of layout	11	29	23	33	39	20
The building itself	6	2	0	1	1	7
Variety, so much to see	24	6	15	5	10	19
Tidy, bright, clean	5	19	10	7	18	48
Movement	2	29	13	2	3	1
Eye-catching	6	10	3	24	5	3
Sense of excitement	3	0	1	10	0	1
Restfulness	0	0	1	0	2	0
Contents	25	17	25	21	13	11
Information provided	7	18	6	14	15	2
Other	8	8	15	17	11	19
Base*	226	194	164	221	226	247

* Visitors who found the gallery very or fairly attractive.

The percentages add to over 100% as people could give more than one answer.

Source of table: the quota samples interviewed as they left selected galleries.

and in the case of the Time Measurement gallery whose unclutteredness was praised by 18 per cent of its visitors.

We again see the importance of spaciousness, or rather of its absence, when we turn to the comments of visitors who found the galleries unattractive (Table 6.9) and improvements which visitors thought could be made to the presentation of each gallery (Table 6.10). The lack of space was the most common complaint of visitors who did not like the appearance of the Aeronautics gallery, and no less than 37 per cent of the visitors who thought the gallery's lay-out would be improved wanted more room to see the exhibits. However, only three per cent suggested achieving this by reducing the number of exhibits.

A considerable proportion of the Exploration gallery's visitors seem to have found it cramped also, and a few of them would have preferred there to be more light. Requests for more working exhibits or for models which could be operated by pressing buttons were common in the Printing and Paper gallery and in the Balcony at the National Railway Museum. Perhaps it was the fact that some of the machines in the Printing and Paper gallery were working already that made its visitors realise that they would like to see some of the others in operation as well. In the Main Hall at the National Railway Museum a similar proportion of visitors suggested that visitors should be allowed to climb on or go inside some of the engines and carriages.

A number of visitors would have liked more help finding their way round individual galleries—particularly the Printing and Paper, Time Measurement and Exploration galleries. Suggestions included notices carrying plans of the gallery concerned or some indication of the order in which one should view the exhibits.

No single factor explains the low rating for attractiveness given to the Time Measurement gallery. The fig-

Table 6.9   Reasons for finding each gallery unattractive at the Science and National Railway Museums

Reasons for finding the gallery unattractive	Science Museum				National Railway Museum	
	Aeronautics	Printing and Paper	Time Measurement	Exploration	Balcony	Main Hall
	%	%	%	%	%	%
Poor lighting	4	0	7	29	0	19
(Lack of) colour	0	17	14	5	10	13
Cluttered, cramped	32	13	16	29	19	13
Too much to see	8	13	7	19	5	0
Glass cases	0	0	5	0	24	0
Lack of sequence, no planned route	16	17	11	10	10	19
Miscellaneous aspects of lay-out	24	17	18	19	33	38
Impersonal atmosphere, dingy, dull	8	17	23	14	14	25
Not eye-catching	4	4	18	0	14	25
Content	28	29	23	14	19	25
Other	28	29	26	24	52	19
Base*	27	24	58	21	22	16

* Visitors who found the gallery not really attractive.
The percentages add to over 100% as people could give more than one answer.
Source of table: the quota samples interviewed as they left selected galleries.

Table 6.10   Improvements suggested to the presentation of each gallery at the Science and National Railway Museums

Improvements suggested	Science Museum				National Railway Museum	
	Aeronautics	Printing and Paper	Time Measurement	Exploration	Balcony	Main Hall
	%	%	%	%	%	%
Lighting	16	4	6	13	6	12
Colour	4	0	9	3	5	2
More space	37	21	16	25	10	17
More realism	6	19	0	2	13	7
Design of cases	0	0	0	0	1	0
(Chrono)logical sequence	7	7	12	4	5	6
Miscellaneous improvements to lay-out	30	12	22	22	25	19
Fewer exhibits	3	1	2	2	4	2
Different exhibits	11	10	14	6	11	6
More working exhibits	7	22	10	7	25	10
Exhibits to climb on, go inside	7	0	0	1	0	25
Information provided	16	22	27	15	14	7
Help finding way round	8	18	22	19	10	12
Other	13	15	19	18	18	21
Base*	90	69	86	90	84	84

* Visitors who thought the presentation could be improved.
The percentages add to over 100% as people could give more than one answer.
Source of table: the quota samples interviewed as they left selected galleries.

ures in Table 6.9 suggest that a dull atmosphere, lack of colour and a feeling on the part of some visitors that the gallery was too cluttered helped to reduce its appeal. A few visitors complained that it was not eye-catching which suggests that part of the reason might be the absence of obvious 'special effects'.

Drawing together the results for the galleries at the two technical museums we find that, as in the Victoria and Albert Museum, visitors thought it important to have enough space to see the exhibits properly. Lighting was also important though less so than at the Victoria and Albert Museum. 'Special effects' however seem to have been much more important—particularly effects which somehow added to the 'realism' of the objects on view.

c. Explaining the exhibits
i. The Victoria and Albert Museum
Information in the four Victoria and Albert Museum galleries was provided by means of labels and printed explanations. In general the text was fairly brief. Table 6.11 shows that over 80 per cent of the visitors interviewed on leaving each gallery had looked at some at least of the printed information provided. Indeed the figures given in Table 6.12 suggest a considerable concern for information. Between a third and a half of each gallery's visitors would have welcomed more information and rather more felt that the way the information was given could be improved.

Most of the additional information wanted was of a kind which would have helped to set the exhibits in context. Many of the suggestions recorded in Table 6.13 were for more information about the origins and history of the exhibit itself or for background information about the times in which it was made. As one visitor said, *"So much of the history has to be incorporated to understand. The whole lifestyle is incorporated in the work"*.

The desire to understand the context of the work seems also to underline the frequent requests for information on the purposes for which items were originally made as well as the requests in the British Sculpture gallery for more information about the subjects depicted in the various works. It cannot, apparently, be assumed that all visitors share equally the basic knowledge and cultural assumptions which help to give these works of art their meaning.

A more technical interest is reflected in the requests for more details of how the exhibits were made and of the materials that were used. In the British Sculpture and Continental Seventeenth Century Art galleries there were frequent requests for more information about the people who made the items.

A number of visitors jumped ahead at this question and referred to ways in which the presentation of information might be improved. Table 6.14 incorporates their comments along with those of the visitors who offered their suggestions in response to the specific question *"Do you feel that the way information is presented (in the gallery concerned) could be improved?"*

Table 6.11 Use of labels and printed explanations in galleries at the Victoria and Albert Museum

	Art of China and Japan	British Sculpture	Tudor Art	Continental 17th Century Art
	%	%	%	%
Read labels or explanations	83	84	84	87
Did not read labels or explanations	17	16	16	13
Base	232	240	251	236

Source of table: the quota samples interviewed as they left selected galleries.

Table 6.12 Percentages of visitors to the Victoria and Albert Museum galleries who would have liked more information or who felt that improvements were possible in the way the information was presented

	Art of China and Japan	British Sculpture	Tudor Art	Continental 17th Century Art
Would like more information	51%	38%	36%	46%
Thought the presentation of information could be improved	54%	42%	55%	58%
*Base**	192	201	210	205

* Excludes visitors who did not read any of the labels or explanations.
Source of table: the quota samples interviewed as they left selected galleries.

Table 6.13 Further information wanted by visitors to the Victoria and Albert Museum galleries

Information wanted	Art of China and Japan	British Sculpture	Tudor Art	Continental 17th Century Art
	%	%	%	%
More detail, technical information	9	11	15	8
How things work	0	0	1	0
How things were made, materials used	29	12	24	19
Uses, purpose	22	21	19	26
Who made things	5	33	8	24
Origins, history of things	41	31	28	39
Historical background	32	24	35	18
Information about the subject depicted	1	23	8	5
Better presented information	26	16	37	28
Don't know	1	0	3	3
Other	19	12	13	14
Base	98	75	75	95

The percentages add to over 100% as people could give more than one answer.
Source of table: the quota samples interviewed as they left selected galleries.

Table 6.14 Improvements suggested in the way information is presented in the Victoria and Albert Museum galleries

Improvements suggested	Art of China and Japan	British Sculpture	Tudor Art	Continental 17th Century Art
	%	%	%	%
Larger print	37	33	45	28
Other improvements to labels	26	29	46	20
More or better positioned labels	24	41	26	47
More detailed descriptions	18	11	18	20
Simpler descriptions	5	4	7	2
Different grades of information	4	11	4	3
Other information sources, such as telephones, slide displays etc	2	1	4	5
Leaflets	9	7	5	10
Illustrations, photographs	3	0	8	6
Other	21	20	16	11
Base	104	85	115	118

The percentages add to over 100% as people could give more than one answer.

Source of table: the quota samples interviewed as they left selected galleries.

Numerous visitors suggested using larger print in the labels, or making various other improvements—principally making them larger, improving the lighting on them or generally brightening them up. The gallery whose labels were felt to need these improvements most was the one devoted to Tudor Art. Another frequent complaint was that there were not enough labels or that, if they were there, they were hard to find—points which were raised particularly often in the British Sculpture and Continental Seventeenth Century Art galleries.

These were the main ways in which visitors felt that improvements could be made. A number, referring back to the theme of the previous question, asked for more detailed information and a few asked for the information to be provided in a simpler form or suggested that it should be clearly graded so that individual visitors could easily pick out the information that would interest them.

A few visitors suggested providing more leaflets. (Some leaflets were in fact already available.) Hardly anyone suggested using any audio-visual devices such as telephones or slide displays to supplement the written information in the galleries.

ii. The Science and National Railway Museums

Four of the six galleries studied at the Science and National Railway Museums made use of some kind of audio-visual equipment to help convey information. Time Measurement and the Main Hall at the National Railway Museum were the two galleries which stuck entirely to printed information. Interestingly these are the two galleries which, according to Table 6.15, had the highest percentages of visitors who made no use at all of the information provided. It seems that the loudspeakers, slide shows, televisions and telephones succeeded in interesting some visitors who would not have been willing to extract the information from labels.

It is worth noting here that the percentages of visitors using different sources of information, or not using any of them, would probably have been different had we included children aged ten or less in the samples whom we interviewed in the galleries. Our results here relate solely to visitors aged 11 or over.

It is clear from Table 6.15 that the audio-visual information sources do not so much replace labels as supplement them. The proportion of visitors who made some use of the printed information is just as high—between 73 and 82 per cent—in those galleries which provided other sources of information as in the two which did not.

Television films were next in terms of the numbers using them in the Exploration gallery, the only one of the galleries we studied which had them. Fifty-six per cent of the Printing and Paper gallery's visitors had listened to one or more of the loudspeaker recordings (although 61 per cent of the Exploration gallery's visitors reported listening to loudspeaker recordings this was often probably just another way of saying that they had looked at the television films to which most of the recordings were linked). Fifty per cent of visitors to the Balcony at the National Railway Museum

Table 6.15 Use of information sources in galleries at the Science and National Railway Museums

Visitors using:	Science Museum				National Railway Museum	
	Aeronautics	Printing and Paper	Time Measurement	Exploration	Balcony	Main Hall
	%	%	%	%	%	%
Labels and printed explanations	76	73	74	82	79	81
Telephone recordings	34	—	—	42	—	—
Loudspeaker recordings	—	56	—	61	—	—
Television films	—	—	—	69	—	—
Slide shows	—	28	—	22	31	—
Tape and slide theatre	—	—	—	—	50	—
Percentage not using any information source	16	12	26	6	9	19
Base	255	222	227	243	248	263

Source of table: the quota samples interviewed as they left selected galleries.

had watched some of the slides and listened to the accompanying commentary in the Tape and Slide Theatre. This is more than looked at the small slide displays in the same gallery or in the Printing and Paper gallery at the Science Museum. Twenty-two per cent of the Exploration gallery's visitors had looked at one of the slide shows there—the most prominent of which was the 'Story of Climate' slide show which occupies what is in effect a tape and slide theatre. The telephone recordings were used by a third of the visitors to the Aeronautics gallery and by a slightly higher proportion in the Exploration gallery.

Table 6.16, which refers only to the galleries with more than one information source, sets out the proportions of visitors preferring each medium. Visitors who used only one source were counted as preferring it, visitors who used more than one were asked which they found best. Print was the most popular medium in the Aeronautics and Printing and Paper galleries as well as in the Balcony at the National Railway Museum. In all three galleries it received the votes of over half the visitors. In the Exploration gallery however print received only 30 per cent of the votes, television films being if anything slightly more popular.

The loudspeaker and telephone recordings were also fairly popular. The percentages in Table 6.16 are of course based on all the information users in each gallery. If we bear in mind that less than half the visitors to the relevant galleries actually used the telephones, their popularity among those who did is impressive. The Tape and Slide Theatre in the Balcony at the National Railway Museum was also very popular among those who sat down to watch. Otherwise the various slide displays seem to have been less popular than other sources—in large part because, as Table 6.15 showed, they failed to attract as much attention.

Table 6.17 sets out the reasons visitors to the Exploration gallery gave for their preferences. The pattern in the other three galleries was very similar. Print was praised as a medium from which information could be extracted very quickly, for allowing visitors to go at their own pace and because visitors could pick out just

Table 6.16 Information source preferred in galleries with more than one information source at the Science and National Railway Museums

Information source preferred	Science Museum			National Railway Museum Balcony
	Aero-nautics	Printing and Paper	Explo-ration	
	%	%	%	%
Labels and printed explanations	75	55	30	57
Telephone recordings	24	—	14	—
Loudspeaker recordings	—	28	10	—
Television films	—	—	33	—
Slide shows	—	14	6	6
Tape and slide theatre	—	—	—	34
Don't know	1	3	6	3
Base*	212	190	223	218

Excludes visitors who did not use any information source.
Source of table: the quota samples interviewed as they left selected galleries.

the information that interested them. The great strength of printed information is, in short, that it allows the visitor to retain his autonomy unlike the visitors using the other media who are forced to go at the pace of the machine.

The other media have two compensating advantages. The first is that those who prefer them find them easier than print and are therefore able to take in more information. The second is that they enable people to see and hear things at the same time. This point deserves some discussion. Some of the visitors who made this point were referring to the fact that the visitor who is getting information from a telephone or loudspeaker recording can look at the exhibit at the same time, something which is not possible when reading a label. As one visitor remarked, *"You could see it as they were talking about it"*.

However the visitors who particularly liked the television films cannot have meant this when 64 per cent of them gave being able to watch and hear at the same time as the reason for their choice. What they were watching was not the exhibit but the film. The same

Table 6.17 Reasons given for preferring different information sources in the Exploration gallery at the Science Museum

Reason given	Information source preferred				
	Labels and printed explanations	Telephone recordings	Loudspeaker recordings	TV films	Slide shows
	%	%	%	%	%
Quick	39	3	9	3	0
Can go at own pace	39	0	0	0	8
Can pick out relevant information	16	0	0	0	0
Easier	16	42	57	26	46
Gives more information	16	48	13	47	54
Interesting	0	3	0	5	8
Can see and hear	0	10	26	64	38
Only source noticed	3	0	0	1	0
Other	19	35	30	19	23
Base*	67	31	23	74	13

Excludes visitors who did not use any information source.
The percentages add to over 100% as people could give more than one answer.
Source of table: the quota sample interviewed as they left the gallery.

applies to the visitors who praised some of the slide shows for combining sight and sound. Many visitors seem simply to prefer using both senses at once. Possibly they find it difficult to concentrate all their attention on the purely visual activities of looking or reading. This may in fact be the reason many visitors find it easier to absorb information from the audio-visual media.

Table 6.18 sets out the proportions of visitors to each of the galleries studied at the two technical museums who would have liked additional information or saw room for improvement in the way it was presented. The scope for change was felt to be greatest in the Science Museum galleries, though the proportions suggesting more or better presented information were lower than in the Victoria and Albert Museum.

Table 6.18 Percentages of visitors to the galleries at the Science and National Railway Museums who would have liked more information or who felt that improvements were possible in the way the information was presented

	Science Museum				National Railway Museum	
	Aeronautics	Printing and Paper	Time Measurement	Exploration	Balcony	Main Hall
Would like more information	26%	26%	23%	28%	13%	18%
Thought the presentation of information could be improved	37%	33%	40%	32%	22%	31%
base*	*215*	*196*	*169*	*227*	*227*	*212*

* Excludes visitors who did not use any information source.
Source of table: the quota samples interviewed as they left selected galleries.

Table 6.19 Further information wanted by visitors to the galleries at the Science and National Railway Museums

Information wanted	Science Museum				National Railway Museum	
	Aeronautics	Printing and Paper	Time Measurement	Exploration	Balcony	Main Hall
	%	%	%	%	%	%
More detail, technical information	15	20	8	38	27	18
How things work	13	33	39	11	3	10
How things were made, materials used	5	6	6	8	0	5
Uses, purpose	16	14	6	5	3	5
Who made things	13	8	3	5	10	3
Origins, history of things	42	18	14	8	27	36
Historical background	7	10	6	0	17	3
Better presented information	56	55	42	60	47	51
Don't know	7	8	6	6	10	0
Other	15	12	17	13	17	8
Base	*55*	*51*	*38*	*64*	*30*	*39*

The percentages add to over 100% as people could give more than one answer.
Source of table: the quota samples interviewed as they left selected galleries.

Table 6.20 Improvements suggested in the way information is presented in the galleries at the Science and National Railway Museums

Improvements suggested	Science Museum				National Railway Museum	
	Aeronautics	Printing and Paper	Time Measurement	Exploration	Balcony	Main Hall
	%	%	%	%	%	%
Larger print	25	13	38	1	14	6
Other improvements to labels	16	9	16	1	6	6
More or better positioned labels	25	20	18	3	10	32
More detailed descriptions	29	28	10	39	22	20
Simpler descriptions	14	14	15	11	6	25
Different grades of information	4	11	13	3	4	11
(More) telephones, slide displays etc	14	22	13	15	16	3
Leaflets	3	3	4	13	8	0
Illustrations, photographs	10	8	10	13	2	14
Other	24	30	22	27	31	17
Base	*79*	*64*	*68*	*73*	*50*	*66*

The percentages add to over 100% as people could give more than one answer.
Source of table: the quota samples interviewed as they left selected galleries.

The kind of things visitors wanted to be told more about can be seen in Table 6.19. As one would expect the pattern of answers differs considerably from that of visitors to the Victoria and Albert Museum. There was some demand, particularly in the Exploration gallery, for more detailed technical information.

Visitors to the Printing and Paper and Time Measurement galleries asked specifically for more information about how things worked. In contrast to the Victoria and Albert Museum there was little demand for more information about general historical background, though in the Aeronautics gallery and the National Railway Museum a number of visitors wanted to know more about the history of the individual items represented.

However about half the visitors who initially said that they would like more information seem to have decided when they were asked what else they wanted to know that their real concern was less with the amount of information than with the way it was presented. One has the impression that they were dissatisfied with the information in the gallery without quite knowing why. Their suggestions and those of people whose opinions were given in response to the specific question *"Do you feel the way information is presented in* [the gallery concerned] *could be improved?"* are set out in Table 6.20.

Requests for larger print were frequent in the Time Measurement and Aeronautics galleries and more or better positioned labels were commonly suggested in several galleries. The most widespread comment was that the information provided should go into more detail, though most galleries also had a substantial group of visitors who wanted simpler information. A number suggested that the information should be divided into distinct grades. Several visitors to each gallery except the Main Hall at the National Railway Museum suggested making more use of audio-visual presentation and some suggested using more illustrations.

These answers do not suggest any clear consensus on the part of Science Museum and National Railway Museum visitors about how information might be better presented. Indeed, what dissatisfaction there is with the information provided does not seem to be strongly focused either on specific aspects of present-ation or on the lack of particular kinds of information. It seems possible that, for many of the visitors in question, the real source of dissatisfaction is a more general feeling that they cannot quite get to grips with the galleries' subject matter.

It is interesting that not many visitors volunteered the suggestion that leaflets should be provided, since it casts considerable doubt on the answers we received to a follow up question that was asked only in the galleries at the Science Museum. Visitors who thought that there was room for improvement in the gallery's presentation

Table 6.21 Whether visitors to the Science Museum galleries would be interested in leaflets if they were provided in the gallery

	Aeronautics	Printing and Paper	Time Measure-ment	Exploration
	%	%	%	%
Would be interested	83	64	83	78
Would not be interested	17	36	17	22
Base*	79	64	68	73†

* *The question on which this table is based was put to people who thought that the way information was provided could be improved.*

† *Twenty-seven of these 73 visitors did not provide an answer to this question. A possible reason for this exceptionally high proportion of 'no answers' is that leaflets describing the Exploration gallery were available from the museum's information office nearby. Some people may therefore have thought that the question did not apply.*

Source of table: the quota samples interviewed as they left selected galleries.

of information were asked *"If there were leaflets pro-vided in* (the gallery concerned) *which you could take away and read, would you personally be interested in them?"* The answers set out in Table 6.21 are overwhelmingly positive. However the fact that so few visitors raised the idea spontaneously shows that it was not one which had generally been given any thought. It seems likely that many visitors were merely indicating casual agreement that the idea might be worth a try.

d. The role of information

The previous section discussed visitors' views on the information in each gallery and the way it was pro-vided. The purpose of the present section is to throw light on how the information affects the visitors' interest in each gallery.

We approached this topic by asking visitors to each gallery whether the information had been pitched at the right level from their point of view. Specifically, visitors were asked:

"Whether people find information interesting can depend on how much they know already. Do you feel that the information in [the gallery concerned] *is meant for someone like yourself, or is it meant for someone who knows less than you do or for someone who already knows more than you do?"*

Their answers are set out in Table 6.22. A majority of each gallery's visitors felt that the information was meant for people like themselves.

Table 6.23 shows how visitors' feelings about the level at which each gallery's information is pitched relate to their interest in the gallery. Two points stand out. The first is that visitors who felt that the gallery's information was meant for people like themselves, in general found each gallery more interesting than did people who felt the information was aimed, over their own heads, at people who knew more. The second point is that people who thought the information rather elementary, in that it seemed intended for people who

Table 6.22 Visitors' judgements of whom the information was meant for by gallery

Gallery		Information meant for:				Base*
		People like themselves	People who knew less than they did	People who knew more than they did	Don't know	
Victoria and Albert Museum						
Art of China and Japan	%	62	8	28	2	192
British Sculpture	%	60	5	30	5	201
Tudor Art	%	65	11	18	5	210
Continental Seventeenth Century Art	%	68	5	23	3	205
Science Museum						
Aeronautics	%	62	23	11	4	215
Printing and Paper	%	48	15	30	7	196
Time Measurement	%	69	11	18	2	169
Exploration	%	66	19	11	4	227
National Railway Museum						
Balcony	%	66	19	14	2	227
Main Hall	%	61	16	23	0	212

* *Excludes visitors who did not use any information source.*
Source of table: the quota samples interviewed as they left selected galleries.

Table 6.23 Percentage of visitors rating each gallery 'very interesting' by their judgements of whom the information was meant for

Gallery	Information meant for:			Bases*		
	People like themselves (a)	People who knew less than they did (b)	People who knew more than they did (c)	(a)	(b)	(c)
	% saying 'very interesting'					
Victoria and Albert Museum						
Art of China and Japan	59%	63%	40%	118	16	53
British Sculpture	53%	40%	22%	119	10	60
Tudor Art	52%	58%	34%	137	24	38
Continental 17th Century Art	52%	50%	48%	140	10	48
Science Museum						
Aeronautics	66%	69%	48%	133	49	23
Printing and Paper	34%	52%	26%	93	29	58
Time Measurement	34%	61%	42%	116	18	31
Exploration	66%	65%	56%	150	43	26
National Railway Museum						
Balcony	55%	67%	35%	149	43	31
Main Hall	78%	59%	58%	129	34	48

* *Excludes visitors who did not use any information source.*
Source of table: the quota samples interviewed as they left selected galleries.

knew less than they did, on the whole found the galleries just as interesting as the people who thought the information was pitched at their own level. In other words visitors seem to have difficulty sustaining their interest in galleries where the information is too advanced for them but not in galleries where the information is too elementary.

This pattern is very much what one would have expected in the Victoria and Albert Museum where, as we saw in the previous section, the additional information in greatest demand was the kind of background material that would set the exhibits in context. Well-informed visitors, able to supply this context for themselves, would not be concerned at the lack of appropriate explanatory material but visitors without the same background knowledge would find it a serious handicap. The fact that the same pattern occurs in the two technical museums suggests that background knowledge is equally important for their visitors though, of course, as we saw in the previous section, the kind of

information that is felt to be relevant at the technical museums differs from that at the Victoria and Albert Museum.

The museums' visitors differed also in the ease with which they thought the museum in question could remedy their lack of background knowledge. Table 6.24 shows that visitors to the galleries at the Victoria and Albert Museum who thought that the information was either too advanced or too elementary were very much more likely than other visitors to suggest that more information should be provided. This was in fact the main remedy proposed though, as the figures in Table 6.25 indicate, people who felt the information was pitched at the wrong level for themselves were also rather more likely than other visitors to suggest improvements in the way it was conveyed.

Those visitors to the galleries at the Science and National Railway Museums who felt that the information was pitched at the wrong level resembled their counterparts at the Victoria and Albert Museum in being readier than others to suggest improvements in the way information was presented. They were also more likely than other visitors to ask for additional information to be provided. However their desire for additional information is far less marked than at the Victoria and Albert Museum. It may be that some of the less knowledgeable visitors to the two technical museums felt that if the information were simplified to a level which they could understand, it would no longer do justice to the real complexity of the exhibits. Technically knowledgeable visitors may not have thought that the complexities of which they were aware could be condensed sufficiently to be presented in a way that could easily be understood.

Table 6.24 Percentage of visitors who would have liked more information by their judgements of whom the information was meant for

Gallery	Information meant for:			Bases*		
	People like themselves (a)	People who knew less than they did (b)	People who knew more than they did (c)	(a)	(b)	(c)
	% who would have liked more information					
Victoria and Albert Museum						
Art of China and Japan	38%	67%	75%	118	16	53
British Sculpture	24%	50%	65%	119	10	60
Tudor Art	25%	58%	61%	137	24	38
Continental 17th Century Art	34%	80%	75%	140	10	48
Science Museum						
Aeronautics	19%	41%	41%	133	49	23
Printing and Paper	22%	45%	24%	93	29	58
Time Measurement	18%	11%	45%	116	18	31
Exploration	25%	29%	35%	150	43	26
National Railway Museum						
Balcony	7%	26%	26%	149	43	31
Main Hall	11%	35%	25%	129	34	48

* Excludes visitors who did not use any information source.
Source of table: the quota samples interviewed as they left selected galleries.

Table 6.25 Percentage of visitors who felt that improvements were possible in the way the information was presented by their judgements of whom the information was meant for

Gallery	Information meant for:			Bases*		
	People like themselves (a)	People who knew less than they did (b)	People who knew more than they did (c)	(a)	(b)	(c)
	% who felt that improvements were possible in the way that information was presented					
Victoria and Albert Museum						
Art of China and Japan	49%	75%	62%	118	16	53
British Sculpture	38%	50%	50%	119	10	60
Tudor Art	52%	58%	68%	137	24	38
Continental 17th Century Art	56%	50%	69%	140	10	48
Science Museum						
Aeronautics	32%	49%	50%	133	49	23
Printing and paper	31%	38%	34%	93	29	58
Time Measurement	40%	28%	45%	116	18	31
Exploration	29%	30%	46%	150	43	26
National Railway Museum						
Balcony	17%	26%	39%	149	43	31
Main Hall	24%	38%	44%	129	34	48

* Excludes visitors who did not use any information source.
Source of table: the quota samples interviewed as they left selected galleries.

7 How visitors with different characteristics reacted to the ten selected galleries

a. General

The previous two chapters have discussed visitors' reactions to the exhibits and methods of presentation in selected galleries. This chapter will look at the relationship between visitors' reactions and their age, sex, educational background, whether they were visiting alone, and their views on the importance of learning things from a museum visit. Two related themes will run through the discussion. First there is the possibility that different types of visitor may approach the business of looking round museum galleries in different ways. Secondly it will be interesting to look for methods of presentation which appeal particularly to certain kinds of visitor and which might be useful if it was intended to angle the presentation of any gallery specifically to those visitors.

b. Age

Table 7.1 shows that older visitors appear to find their visits to the galleries particularly interesting. In each of the ten galleries selected for special study (with the partial exception of the British Sculpture gallery) the proportion of visitors reporting that they found the gallery very interesting rises with age. The Art of China and Japan gallery, where the proportion of very interested visitors rises from 38 per cent of the 11–20 age group to 71 per cent among the over-forties, is typical in this respect. There are no comparable figures for interest levels among the under-eleven age group since they were considered too young to be included in the quotas. Chapter 11 contains a section based on what parents told us about the reactions of their young children.

It is possible that the increase of interest levels with age results from some kind of maturational process—as people grow older they might gradually learn how to look at museums and come to be more appreciative of what they have to offer. However the figures in Table 7.1 are not sufficient to demonstrate this. An equally plausible explanation would be that as people grow older those who continue to visit museums are the people who, even when young, enjoyed them most. In that case the increase in interest levels might reflect changes with age in the composition of the visiting public rather than any evolution in the tastes of individual visitors.

Age is also related to the way visitors like to get their information. Looking at Table 7.2 one can see that visitors in the 11–20 age group are marked off from older visitors by their enthusiasm for audio-visual media. Older visitors were much more likely to prefer print. That is not to say, however, that older visitors positively rejected information provided by televisions, loudspeakers and the rest. Their reactions seem to have depended on the use made of print on the one hand and audio-visual media on the other in the particular gallery concerned. In the Science Museum's Exploration gallery the proportion who definitely preferred print was only 48 per cent even amongst the oldest of our four age groups (see Figure 7.1).

The popularity of audio-visual presentation with young visitors may well be due in part to the fact that they, unlike many of the older visitors, belong to a generation for whom television had always been an accepted part of life. They may never have acquired the idea that

Table 7.1 The percentage of visitors ratings each gallery 'very interesting' by age

	Age (years)				Bases			
	11–20 (a)	21–30 (b)	31–40 (c)	41 or over (d)	(a)	(b)	(c)	(d)
Victoria and Albert Museum	*% saying 'very interesting'*							
Art of China and Japan	38%	39%	51%	71%	61	49	43	79
British Sculpture	30%	32%	19%	52%	64	54	31	91
Tudor Art	34%	40%	40%	56%	67	67	30	87
Continental 17th Century Art	33%	41%	45%	73%	69	56	33	78
Science Museum								
Aeronautics	46%	54%	81%	81%	87	76	34	58
Printing and Paper	24%	30%	45%	45%	85	64	33	40
Time Measurement	21%	30%	42%	51%	71	71	38	47
Exploration	50%	59%	78%	89%	79	82	45	37
National Railway Museum								
Balcony	42%	46%	54%	78%	66	91	41	50
Main Hall	51%	62%	81%	83%	74	93	42	54

Source of table: the quota samples interviewed as they left selected galleries.

Table 7.2 Percentages of visitors preferring different sources of information by age

Gallery/source of information	Age (years)			
	11–20	21–30	31–40	41 or over
	%	%	%	%
Science Museum				
Aeronautics				
Printed information	54	91	77	84
Telephone recordings	44	8	23	16
Don't know	3	2	0	0
*Base**	*73*	*64*	*30*	*45*
Printing and Paper				
Printed information	33	68	64	74
Loudspeaker recordings	42	18	24	17
Slide shows	23	11	8	6
Don't know	1	4	4	3
*Base**	*74*	*56*	*25*	*35*
Exploration				
Printed information	17	34	34	48
Telephone recordings	22	11	12	3
Loudspeaker recordings	8	11	12	14
Television	41	37	24	17
Slide shows	11	5	0	3
Don't know	1	3	17	14
*Base**	*76*	*76*	*41*	*30*
National Railway Museum				
Balcony				
Printed information	34	67	61	67
Tape and slide theatre	54	23	33	31
Small slide displays	13	5	0	2
Don't know	0	6	6	0
*Base**	*56*	*84*	*36*	*42*

*Excludes visitors who did not use any of the information sources.

Source of table: the quota samples interviewed as they left selected galleries.

print is the most natural medium for information. It may even be that regular television watching has shaped their expectations passively so that they are less ready than were their elders at the same age to get their information the hard way by reading. However nothing in this survey could show that this was so. It would be equally plausible to maintain that the way people like to obtain information changes with age as part of a progressive alteration in their style of attention.

The major advantage of print, as we saw in Chapter 6, is the autonomy it gives its readers who are free to read as much or as little as they want and to scan the labels to find the points which most interest them. If older visitors did tend to favour a generally autonomous style of attention, it would not be surprising if, as the general drift of the figures in Table 7.3 tends to suggest, they were generally somewhat readier than younger visitors to praise galleries for spaciousness or lack of clutter. These are factors which make it possible for visitors to select for themselves the angle from which to view each exhibit. A self-directed approach to viewing would be a considerable help in many of the galleries we have studied and if it was more common among older visitors might go a long way towards explaining the greater interest that they express.

The figures in Table 7.4—though like those in the previous table they are not entirely clear cut—indicate that younger visitors are more likely than their elders to cite the colours in a gallery as a reason for finding it attractive. This tendency is particularly noticeable at the two technical museums. It is not simply a question of young people preferring bright colours: the comparatively subdued blues and greys of the Aeronautics gallery also received most favourable comments from the gallery's younger visitors. Rather it seems that younger visitors are generally more sensitive to colour. The two Science Museum galleries whose colour schemes appeal to them most, the Exploration and Printing and Paper galleries, also appear to be the two whose colour schemes are most praised by the older visitors—though in both cases it is the younger people who seem to have noticed the colours particularly. The varied colours in the Exploration gallery seem if anything to have been more appealing than the striking contrasts of red, green and white in the Printing and Paper gallery. The variety of colours used to paint the locomotives and rolling stock of the Main Hall at the National Railway Museum were mentioned by 22 per cent of the 11–20s who liked the hall's appearance.

It is tempting to suggest a link between young visitors' greater sensitivity to colour and their preference for audio-visual presentation. We noted in the previous section that one of the main reasons given for prefer-

Figure 7.1 Preferences for different information sources in the Exploration Gallery (from Table 7.2)

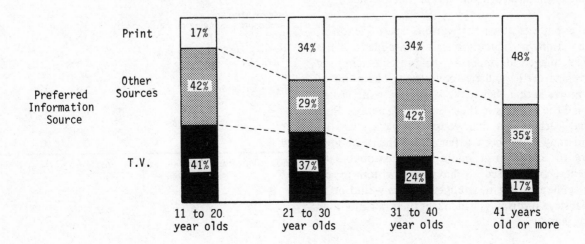

Table 7.3 Percentages of those visitors commenting favourably on each gallery's appearance who praised the spaciousness or lack of clutter, by age

Gallery	Age (years)				Bases			
	11–20 (a)	21–30 (b)	31–40 (c)	41 or over (d)	*(a)*	*(b)*	*(c)*	*(d)*
Victoria and Albert Museum	*% who praised the spaciousness or lack of clutter*							
Art of China and Japan	25%	37%	42%	57%	*51*	*38*	*34*	*69*
British Sculpture	33%	33%	56%	46%	*55*	*38*	*25*	*73*
Tudor Art	11%	7%	5%	12%	*54*	*56*	*22*	*66*
Continental 17th Century Art	19%	41%	23%	24%	*64*	*54*	*28*	*72*
Science Museum								
Aeronautics	11%	7%	10%	25%	*74*	*69*	*31*	*52*
Printing and Paper	10%	6%	28%	22%	*72*	*52*	*32*	*38*
Time Measurement	11%	12%	22%	29%	*48*	*48*	*27*	*41*
Exploration	7%	10%	13%	6%	*72*	*74*	*43*	*32*
National Railway Museum								
Balcony	28%	33%	32%	29%	*58*	*81*	*39*	*48*
Main Hall	19%	36%	49%	39%	*67*	*87*	*41*	*52*

Source of table: the quota samples interviewed as they left selected galleries.

Table 7.4 Percentages of those visitors favourably commenting on each gallery's appearance who praised the colour, by age

Gallery	Age (years)				Bases			
	11–20 (a)	21–30 (b)	31–40 (c)	41 or over (d)	*(a)*	*(b)*	*(c)*	*(d)*
Victoria and Albert Museum	*% who praised the colour*							
Art of China and Japan	14%	5%	3%	12%	*51*	*38*	*34*	*69*
British Sculpture	11%	17%	8%	6%	*55*	*38*	*25*	*73*
Tudor Art	15%	7%	10%	8%	*54*	*56*	*22*	*66*
Continental 17th Century Art	22%	9%	23%	10%	*64*	*54*	*28*	*72*
Science Museum								
Aeronautics	11%	4%	6%	2%	*74*	*69*	*31*	*52*
Printing and Paper	18%	13%	9%	8%	*72*	*52*	*32*	*38*
Time Measurement	0%	0%	0%	0%	*48*	*48*	*27*	*41*
Exploration	24%	20%	15%	6%	*72*	*74*	*43*	*32*
National Railway Museum								
Balcony	5%	6%	3%	6%	*58*	*81*	*39*	*48*
Main Hall	22%	13%	0%	10%	*67*	*87*	*41*	*52*

Source of table: the quota samples interviewed as they left selected galleries.

ring the audio-visual media was that they enabled visitors to see and hear at the same time. The essential point appeared to be that they engaged two senses, not just one. If young people's awareness of colour was an aspect of a greater alertness to sense impressions generally, it would be understandable that they should find it hard to shut themselves off from what they heard going on about them in order to concentrate all their attention on looking at the printed information. Media that occupied their hearing as well as their sight would make the task of assimilation much easier.

In Chapter 6 it was shown that various 'special features' of presentation, particularly ones which added a sense of realism, played an important role in enhancing the attractiveness of the galleries concerned. The point of these features is that they make it easier for the visitor to imagine the exhibits in their 'real-life' context. Pursuing our earlier idea that young people's styles of attention may have been affected by their life-long exposure to television it might seem reasonable to expect that, being used to having television to do their imagining for them, younger visitors would be in particular need of displays which lent their imaginations a helping hand.

Table 7.5 Percentages of those visitors commenting favourably on each gallery's appearance who praised various special features, by age

Gallery/special feature praised	Age (years)			
	11–20	21–30	31–40	41 or over
Victoria and Albert Museum				
Tudor Art: room settings	17%	25%	30%	25%
Base	*54*	*56*	*22*	*66*
Continental 17th Century Art: room settings	11%	11%	0%	12%
Base	*64*	*54*	*28*	*72*
Science Museum				
Aeronautics: hanging aircraft	42%	34%	39%	25%
Base	*74*	*69*	*31*	*52*
Printing and Paper: working exhibits	28%	27%	28%	32%
Base	*72*	*52*	*32*	*38*
Exploration: darkness	13%	26%	15%	3%
sense of excitement	7%	11%	15%	9%
eye-catching quality	21%	23%	30%	25%
Base	*72*	*74*	*43*	*32*

Source of table: the quota samples interviewed as they left selected galleries.

Table 7.6 Percentages of those visitors commenting favourably on each gallery's appearance who praised it for realism, by age

Gallery	Age (years)				Bases			
	11–20 (a)	21–30 (b)	31–40 (c)	41 or over (d)	(a)	(b)	(c)	(d)
Science Museum	*% who praised the realism*							
Aeronautics	21%	13%	10%	8%	74	69	31	52
Printing and Paper	14%	13%	9%	3%	72	52	32	38
Exploration	8%	9%	23%	16%	72	74	43	32
National Railway Museum								
Main Hall	10%	15%	27%	24%	67	87	41	52

Source of table: the quota samples interviewed as they left selected galleries.

Tables 7.5 and 7.6 do not bear this hypothesis out. Table 7.5 gives the percentages (of those liking each gallery's appearance) who commented favourably on the room settings at the Victoria and Albert Museum and on various special features at the Science Museum. Table 7.6 takes the four galleries most praised for realism at the Science and National Railway Museums and relates the frequency of comments on the galleries' realism to age.

The one clear statement that can be made about the rather confusing picture which emerges is that there seems to be no simple relationship between age and a liking for 'realistic' presentation. Looking at Table 7.6 we see that in two of the galleries—those devoted to Aeronautics and to Printing and Paper—it was the younger members of our samples who gave most praise for realism. In the Exploration gallery and in the Main Hall at the National Railway Museum the opposite was the case.

Turning to Table 7.5 we notice that the Victoria and Albert Museum's room settings appear to appeal equally to 11–30 year olds and to the over 40s. It looks as though the hanging aircraft in the Aeronautics gallery may have appealed most to younger visitors but working exhibits—note that here we are talking about exhibits which operated by themselves, not exhibits worked by pushing buttons—do not appear to have had more appeal for any particular age group, if we are to judge from the responses of visitors in the Printing and Paper gallery. Nor do the darkness, sense of excitement and eye-catching quality of the Exploration gallery. In both the Printing and Paper gallery and the Exploration gallery the appeal of these 'special features' seems to have been as strong for older as for younger visitors.

c. Other visitor characteristics
No clear patterns emerge when visitors' overall interest ratings of the ten selected galleries are related to sex, to the ages at which visitors ended their full-time education or to whether they were visiting alone or with family or friends. However these factors are related to visitors' views on the information provided in the galleries.

Table 7.7 shows that women were a good deal more likely than men to think that the information in the galleries at the two technical museums was meant for

people who knew more than they did. This suggests that the two museums' female visitors had less technical background than male visitors and is consistent with the finding of Chapter 3 (Table 3.6) that women were more likely than men to be visiting these two museums 'altruistically'—in order to accompany someone else rather than for their own interest.

The answers of people selected for quota interviews as they left the National Railway Museum also suggest that its women visitors are generally less interested than men in the technical or mechanical aspects of its subject matter. Table 7.8 shows how men and women rated the locomotives and coaches in the Main Hall. While the locomotives appealed most to men and boys, the coaches were much more interesting to women. The

Table 7.7 Percentages of male and female visitors who thought the information in each gallery was meant for people who knew more than they did themselves

Gallery	Males	Females	Bases* Males	Females
	% who thought information was meant for people who knew more than they did			
Victoria and Albert Museum				
Art of China and Japan	31%	24%	98	94
British Sculpture	32%	28%	101	100
Tudor Art	19%	17%	99	111
Continental 17th Century Art	23%	24%	103	102
Science Museum				
Aeronautics	6%	21%	143	72
Printing and Paper	21%	44%	120	76
Time Measurement	9%	33%	106	63
Exploration	7%	18%	131	96
National Railway Museum				
Balcony	9%	20%	131	96
Main Hall	17%	30%	121	91

*Excludes visitors who did not use any of the information sources.
Source of table: the quota samples interviewed as they left selected galleries.

Table 7.8 Percentages of visitors to the National Railway Museum rating the locomotives and the coaches very interesting by sex

	Males	Females	Bases Males	Females
Locomotives	74%	58%	322	268
Coaches	57%	81%	301	261

The figures are based on the replies of those visitors who had stopped to look at the relevant section of the Main Hall.
Source of table: the quota samples of museum leavers.

coaches may therefore have helped to sustain the interest of women visitors despite the limited attraction which the technical side of railways appears to have for them.

Table 7.9 relates Science Museum and National Railway Museum visitors' preferences for different sources of information to the ages at which they finished full-time education. More educated visitors resemble older visitors in their preference for printed information rather than audio-visual media. For instance, the figures which Table 7.9 gives for the Balcony at the National Railway Museum, show that print was the preferred information source for 81 per cent of visitors who had finished their education after their twenty first birthday but of only 52 per cent of those who left school aged 16 or less. Thirty-seven per cent of the least educated group preferred the Tape and Slide Theatre compared with only 11 per cent of the most educated visitors.

Table 7.9 Percentages of visitors preferring different sources of information by the age at which they had finished their full-time education

Sources of information preferred	Age at which finished education		
	16 or under	17–20	21 or over
Science Museum	%	%	%
Aeronautics			
Printed information	76	89	93
Telephone recordings	24	11	7
Don't know	0	0	0
*Base**	*67*	*35*	*43*
Printing and Paper			
Printed information	56	68	68
Loudspeaker recordings	33	18	18
Slide shows	6	13	9
Don't know	6	0	6
*Base**	*36*	*38*	*34*
Exploration			
Printed information	28	43	40
Telephone recordings	15	6	8
Loudspeaker recordings	15	15	4
Television	26	30	29
Slide shows	6	6	2
Don't know	9	0	17
*Base**	*53*	*47*	*49*
National Railway Museum			
Balcony			
Printed information	52	70	81
Tape and slide theatre	37	23	11
Small slide displays	6	4	3
Don't know	5	2	6
*Base**	*86*	*47*	*36*

*Excludes visitors who did not use any of the information sources. Visitors still in full-time education are also excluded.

Source of table: the quota samples interviewed as they left selected galleries.

Table 7.10 shows that people visiting alone in galleries at the two technical museums are considerably more likely to wish for additional information than people visiting in company. There may be several reasons for this. A proportion of the people who were visiting with others must have been 'altruistic visitors' who were not themselves particularly interested in the contents of the gallery and therefore not particularly eager for extensive information about them. (It was shown in Chapter 3, Section (b), that 'altruistic' visiting is far more common at the two technical museums than at the Victoria and Albert Museum.) In addition as Table 3.9 (Chapter 3) indicated, people visiting in company seem to attach somewhat less importance to learning from a museum visit than solitary visitors do. This might imply less enthusiasm for additional information.

A third factor may be the effect of belonging to a group on each individual's attention span—it can be quite hard to absorb much printed material or attend for long to the audio-visual media and at the same time keep up with, or avoid holding up, the people one is with. Accompanied visitors may have felt that, in the circumstances, they would not have had time to absorb more information. This factor too would have been more important at the Science and National Railway Museums than it would at the Victoria and Albert Museum where the information on the comparatively short labels would have taken less time to absorb.

Table 7.10 Percentages of visitors who would have liked more information by whether they were alone or accompanied

Gallery	Alone (a)	Accompanied (b)	Bases* (a)	(b)
	% who would have liked more information			
Victoria and Albert Museum				
Art of China and Japan	54%	50%	*68*	*116*
British Sculpture	34%	40%	*70*	*124*
Tudor Art	38%	36%	*65*	*136*
Continental 17th Century Art	40%	48%	*77*	*120*
Science Museum				
Aeronautics	43%	20%	*42*	*145*
Printing and Paper	36%	24%	*36*	*130*
Time Measurement	30%	22%	*43*	*115*
Exploration	37%	24%	*51*	*149*
National Railway Museum				
Balcony	20%	11%	*46*	*168*
Main Hall	22%	17%	*46*	*153*

*Excludes visitors who did not use any of the information sources.

Source of table: the quota samples interviewed as they left selected galleries.

Table 7.11 deals with the preferences of solitary and accompanied visitors to the two technical museums for different information sources. The figures for the Aeronautics and Printing and Paper galleries show that, as one might expect, solitary visitors are more likely than others to prefer the autonomy offered by print to the comparative ease of using the audio-visual media. The figures for the Balcony and the Exploration gallery however reflect the solitary visitor's greater freedom to devote a considerable time to one part of the gallery. Thus the Tape and Slide Theatre in the National Railway Museum's Balcony, whose shows run for nearly half an hour, appears to be more popular with the less constrained solitary visitors than with accompanied visitors. The time needed to watch the

Table 7.11 Percentages of visitors preferring different sources of information by whether they were alone or accompanied

Source of information preferred	Alone	Accompanied
Science Museum	%	%
Aeronautics		
Printed information	93	77
Telephone recordings	7	21
Don't know	0	2
*Base**	*42*	*142*
Printing and Paper		
Printed Information	69	56
Loudspeaker recordings	11	27
Slide shows	17	15
Don't know	3	2
*Base**	*36*	*125*
Exploration		
Printed information	35	32
Telephone recordings	6	13
Loudspeaker recordings	2	14
Television	37	29
Slide shows	12	5
Don't know	8	7
*Base**	*51*	*146*
National Railway Museums		
Balcony		
Printed information	43	65
Tape and slide theatre	43	28
Small slide displays	7	5
Don't know	7	2
*Base**	*44*	*161*

*Excludes visitors who did not use any of the information sources.

Source of table: the quota samples interviewed as they left selected galleries.

Table 7.12 Percentage of visitors rating each gallery 'very interesting' by views on the importance of learning

Gallery	Important to feel you have learnt something (a)	Just enjoy looking at things (b)	Base (a)	(b)
	% saying 'very interesting'			
Victoria and Albert Museum				
Art of China and Japan	58%	43%	113	108
British Sculpture	42%	28%	139	95
Tudor Art	54%	29%	147	103
Continental 17th Century Art	55%	42%	142	89
Science Museum				
Aeronautics	69%	48%	141	106
Printing and Paper	35%	27%	143	75
Time Measurement	38%	24%	130	87
Exploration	69%	56%	145	91
National Railway Museum				
Balcony	66%	38%	134	112
Main Hall	74%	58%	128	129

Source of table: the quota samples interviewed as they left selected galleries.

Table 7.13 Percentage of visitors to each gallery who would have liked more information by views on the importance of learning

Gallery	Important to feel you have learnt something (a)	Just like looking at things (b)	Bases* (a)	(b)
	% who would have liked more information			
Victoria and Albert Museum				
Art of China and Japan	52%	50%	96	86
British Sculpture	41%	36%	116	79
Tudor Art	37%	34%	128	82
Continental 17th Century Art	46%	44%	128	72
Science Museum				
Aeronautics	34%	16%	116	91
Printing and Paper	25%	25%	127	65
Time Measurement	28%	14%	104	58
Exploration	29%	26%	135	85
National Railway Museum				
Balcony	16%	10%	126	99
Main Hall	22%	15%	108	99

*Excludes visitors who did not use any of the information sources.

Source of table: the quota samples interviewed as they left selected galleries.

television recordings right through may explain why those in the Exploration gallery appeal to solitary visitors at least as much as to those accompanied by family or friends.

d. Looking and learning

Another factor which is relevant to visitors' reactions to the galleries is the distinction discussed in Chapter 3, Section (c), between visitors who say that it is important 'to feel you have learnt something' from going round a museum and visitors who say that they 'just enjoy looking at things'. Table 7.12 shows that 'learners' reported higher interest levels in every gallery than visitors who simply came to look.

Table 7.13 sheds some light on what 'learners' meant when they said that it was important to feel one had learnt something from a museum visit. It shows that, as might be expected, more 'learners' than 'lookers' would have liked additional information provided in the galleries. However in the Victoria and Albert Museum the difference between the proportions of 'lookers' and 'learners' wanting additional information was very slight. These figures for the Victoria and Albert Museum are particularly noteworthy since only a limited amount of information was provided in the four galleries. The fact that 'learners' expressed very little extra demand for information suggests that when visitors to the Victoria and Albert Museum said that "learning" was important they did not usually mean

that they intended to absorb a great deal of printed information. It seems likely that for them learning was mainly a visual process, a matter of taking seriously the exhibits they were looking at.

At all three museums it may be that what distinguished the people who said they wanted to learn from their visits from those who said they just wanted to look at things was that the 'learners' were more committed in their approach to museum visiting. A plausible interpretation of Table 7.12 would then be that people whose general approach is more committed are more likely to find reasons for interest in any particular gallery.

However there is a tantalising association between age and both the importance given to learning (see Chapter 3, Table 3.7) and the level of interest reported (see Table 7.1). It is possible to take two different views of this set of associations. It may well be that older visitors are more committed in their approach and therefore extract more interest from the galleries they visit. But it is also possible that the relationship between interest levels and commitment to learning is an incidental result of their common relationship to the visitor's age. In that case there would be no direct relationship between interest and commitment to learning.

8 Patterns of visiting

a. General

The last four chapters have been concerned with how visitors set about looking at particular galleries. The theme of this chapter is the way they set about visiting each museum as a whole. The discussion will cover differences between visits and return visits, the extent of each visit and the factors that determine which galleries in each museum attract most visitors. We will go on to discuss the role of special temporary exhibitions in attracting visitors to the two London museums and the reasons for the marked seasonal variation in the daily attendance at the National Railway Museum.

b. Visiting and revisiting

The proportion of visitors who had been to the Victoria and Albert and Science Museums before rises steadily with age. Table 8.1 shows that only 42 per cent of children aged ten or less who were visiting the Science Museum had been there before. On the other hand 70 per cent of the 41 and over age group were making a return visit. This age gradient does not appear in the case of the National Railway Museum. Since the National Railway Museum has only been open since the mid 1970s older visitors have not had any more years in which to visit it than all but the youngest of the younger visitors.

Table 8.2, which is restricted to visitors who had completed their full-time education after continuing it to the age of 17 or beyond, relates the proportion of visitors who were making return visits, to the main subject areas they studied during the last two years of their education. There appears to be no relationship at all at the two technical museums, but at the Victoria and Albert Museum people who studied the visual arts were far more likely than others to be making a return visit. It seems that the Victoria and Albert Museum can offer enough to the specialist visitor to make a

Table 8.2 Proportion of visitors who had visited each museum before by main subjects studied during last two years of full-time education

Museum	Visual arts	Other 'arts' subjects	Science	Other or general studies
	% who had visited before			
Victoria and Albert Museum	82%	55%	38%	58%
*Weighted base**	*130*	*170*	*90*	*220*
Science Museum	—	63%	64%	68%
*Weighted base**	*10*	*80*	*130*	*140*
National Railway Museum	—	34%	31%	35%
*Weighted base**	*20*	*100*	*100*	*140*

* Visitors who completed their education at the age of 17 or over.
Source of table: weighted results from count-based samples of museum leavers.

number of return visits worth while but provides rather less motivation for the non-specialist to return. The Science Museum attracts return visits from slightly more non-specialist visitors but does not attract more return visits from people who had specialised in science than from anybody else. This difference between the two museums is consistent with the finding in Chapter 2 that the Science Museum was more of a 'family museum' than was the Victoria and Albert Museum.

There is a relationship between previous visiting and the company, if any, people arrive with. The results in Table 8.3 demonstrate that solitary visitors are more likely than any other category to have visited each museum before. Two thirds of the Victoria and Albert Museum's solitary visitors, and three quarters of the people visiting the Science Museum alone, had been to the museum concerned before. The proportion who had been to each museum before was lower in the case of people visiting with friends and lower still in the case of people visiting with their families or in organised parties.

The same general relationship holds for younger visitors as is shown by Table 8.4 which is restricted to people aged 20 or under: it indicates that between 60 and 70 per cent of children visiting the museums with their schools are making their first visit.

The need to keep their interviews short prevented us from going on to ask the people selected in the count-based samples who had visited each museum before how often they visited the museum concerned. However we were able to raise this topic with the quota-based samples of people leaving each museum. Tables 8.5 and 8.6 are based on the replies of quota sample visitors who had visited the Victoria and Albert Museum and

Table 8.1 Proportion of visitors who had visited each museum before by age

Museum	Age (years)				
	0–10	11–20	21–30	31–40	41 and over
	% who had visited before				
Victoria and Albert Museum	27%	39%	56%	60%	63%
Weighted base	*60*	*230*	*280*	*160*	*270*
Science Museum	42%	50%	52%	65%	70%
Weighted base	*190*	*360*	*150*	*150*	*140*
National Railway Museum	30%	43%	38%	39%	27%
Weighted base	*110*	*210*	*150*	*270*	*260*

Source of table: weighted results from count-based samples of museum leavers.

Table 8.3 Proportion of visitors who had visited each museum before by whether visiting the museum alone or in a group

Museum	Visiting:				
	Alone	With friends but not members of family	With members of family	With a school party	With another type of organised party
	% who had visited before				
Victoria and Albert Museum	68%	58%	41%	44%	56%
Weighted base	*260*	*240*	*380*	*60*	*60*
Science Museum	76%	58%	57%	35%	41%
Weighted base	*100*	*200*	*450*	*190*	*60*
National Railway Museum	54%	41%	33%	39%	33%
Weighted base	*50*	*110*	*640*	*140*	*60*

Source of table: weighted results from count-based samples of museum leavers.

Table 8.4 Proportion of younger visitors (aged 20 or under) who had visited each museum before by whether visiting the museum alone or in a group

Museum	Visiting:				
	Alone	With friends but not members of family	With members of family	With a school party	With another type of organised party
	% who had visited before				
Victoria and Albert Museum	39%	43%	31%	38%	29%
Weighted base	*40*	*80*	*100*	*50*	*20*
Science Museum	78%	63%	49%	32%	40%
Weighted base	*30*	*110*	*190*	*170*	*40*
National Railway Museum	—	48%	33%	36%	46%
Weighted base	*10*	*40*	*120*	*130*	*30*

Source of table: weighted results from count-based samples of museum leavers.

Table 8.5 Recent visiting by people who had been to the Victoria and Albert Museum before: number of visits during the past 12 months by whether making the present visit alone

Number of visits during the past 12 months	Visiting:			All previous visitors
	Alone	With friends but not members of family	With members of family	
	%	%	%	%
None	31	43	56	41
1	16	24	16	18
2–4	27	20	17	22
5–9	11	6	7	8
10 or more	15	7	5	10
Base	*241*	*138*	*155*	*534*

Source of table: quota samples of museum leavers.

Table 8.6 Recent visiting by people who had been to the Science Museum before: number of visits during the past 12 months by whether making the present visit alone

Number of visits during the past 12 months	Visiting:			All previous visitors
	Alone	With friends but not members of family	With members of family	
	%	%	%	%
None	39	61	65	56
1	22	22	19	21
2–4	21	11	13	15
5–9	5	4	1	3
10 or more	12	3	2	5
Base	*203*	*198*	*265*	*666*

Source of table: quota samples of museum leavers.

the Science Museum before. They relate the number of visits made in the previous 12 months to whether the individual was making the visit alone, with friends or with members of his or her family (party visitors were not included in the quota samples of museum leavers). The results show that not only are people going to the museums alone more likely than others to be past visitors but that solitary past visitors are more likely than other past visitors to have visited the museum during the previous year. In other words people visiting alone are particularly likely to be regular visitors.

(It should be noted that, because the quota samples deliberately over-represented lone visitors, the right-hand columns of Table 8.5 and 8.6 relating to all previous visitors somewhat overstate the proportion who had made a visit in the previous 12 months.)

c. The extent of the visit

Both London museums are very large—in each case our plan shows over 40 permanent galleries—so visitors are faced with a choice between either concentrating their visit on a small part of the museum and looking at it fairly thoroughly or attempting to see more of the museum but in a less concentrated way. This choice is the subject of the present section.

The following procedure was used in the quota sample interviews with museum leavers to establish which galleries each visitor had walked through or stopped to look at. The interviewer and the visitor were both seated by a small table on which were spread the two sheets of the museum plan. Using these as aids the interviewer helped the visitor to reconstruct the order of his or her visit.

First the interviewer asked the informant to indicate the galleries where he or she had stopped to look at something, starting with the first gallery visited and continuing in the order in which the visitor had been to them. The interviewer marked these galleries on the plan. When this was done the interviewer asked the informant to show on the plan how he or she got from one gallery to the next. The interviewer then marked the route on the plan in order to provide a record of the galleries which the visitor had simply walked through. At this stage some visitors remembered other galleries which they had stopped to look at and these were incorporated in the record of the visit. Finally, after the visitor's itinerary had been fully established, the interviewer asked how interesting the visitor had found each gallery where he or she had stopped to look at something. The answers to this last question will be discussed in Chapter 9.

The itineraries we recorded are of course only as reliable as the visitors' memories. However one advantage of actually marking the route on the museum plan was that in the great majority of cases we were able to obtain itineraries which at least appeared to be complete—in that there were no breaks in the reported route. Eighty-two per cent of the visitors interviewed at each museum recorded apparently complete routes.

Almost everyone, however, provided a list of galleries where they had stopped to look at something and Table 8.7 is based on their replies. The figures show that few visitors attempted to look at exhibits in more than a small proportion of each museum's galleries. Thirty-seven per cent of the visitors interviewed at the Victoria and Albert Museum had restricted their viewing to four galleries or less. Only four per cent of the Victoria and Albert Museum's visitors and 12 per cent of the people interviewed at the Science Museum had stopped to look at more than 15 galleries. (There is some indirect evidence, discussed in the first section of Chapter 9, that visitors to the Victoria and Albert Museum may sometimes have forgotten that they stopped to look at

galleries where they only spent a short time. If so the figures quoted here will understate the extent of their visits. However this is unlikely to affect the validity of the general point that most visitors only stopped to look at exhibits in a few of the museum's galleries.)

The number of galleries people stopped to look at depended partly on how familiar they were with the museum in question. People's initial visits to each museum appear to be relatively wide ranging and exploratory—stopping to look at an average of eight galleries in the Victoria and Albert Museum and ten in the Science Museum, according to Table 8.8. People who had visited the museum before restricted themselves to fewer galleries. This was particularly true of people who had already visited the museum twice or more in the past 12 months: the average number of galleries viewed by people who had visited two to four times in the past 12 months was 4.8 at the Victoria and Albert Museum and 5.9 at the Science Museum.

Table 8.9 relates the number of galleries visitors stopped to look at to whether they were visiting alone, with friends, or with members of their families. In both museums accompanied visitors seem to be rather

Table 8.7 The number of galleries visitors stopped to look at in the two London museums

Number of galleries* stopped to look at	Victoria and Albert Museum	Science Museum
	%	%
1–2	18	10
3–4	19	15
5–6	18	18
7–10	26	27
11–15	14	18
16 or more	4	12
Base†	1,001	1,044

* Including the entrance hall and the museum shop.

† Visitors who provided a list of the galleries they stopped to look at.

Source of table: the quota samples of museum leavers.

Table 8.8 The mean numbers of galleries visitors stopped to look at in the two London Museums by previous visiting

	Victoria and Albert Museum	Science Museum	Bases	
	(a)	(b)	(a)	(b)
	Mean no. of galleries*			
Never visited before	8.0	10.0	468	378
Visited before but not in past 12 months	6.4	8.9	217	368
Visited once in past 12 months	5.6	8.2	97	139
Visited 2 to 4 times in past 12 months	4.8	5.9	118	98
Visited 5 to 9 times in past 12 months	4.6	4.9	44	20
Visited 10 or more times in past 12 months	3.8	4.3	52	34
All visitors†	6.7	8.7	1,001	1,044

* Including the entrance hall and the museum shop.

† Visitors who provided a list of the galleries they stopped to look at.

Source of table: the quota samples of museum leavers.

Table 8.9 The mean numbers of galleries visitors stopped to look at in the two London museums by whether they came to the museum alone, with friends or with members of their families

	Victoria and Albert Museum	Science Museum	Bases	
	(a)	(b)	(a)	(b)
	Mean no. of galleries*			
Alone	5.8	6.5	360	287
With friends but not members of family	6.5	9.6	280	320
With members of family	7.8	9.5	361	437
All visitors†	6.7	8.7	1,001	1,044

* Including the entrance hall and the museum shop.

† Visitors who provided a list of the galleries they stopped to look at.

Source of table: the quota samples of museum leavers.

more wide-ranging than solitary visitors. These latter restrict themselves to fewer galleries attempting, perhaps, to study them in greater depth.

So far we have talked about the number of galleries that visitors stopped to look at, leaving aside the galleries that visitors walked through without stopping. It might be that visitors actually ranged over the whole of each museum walking through every gallery but only stopping to look at the few galleries which caught their fancy. In fact this does not seem to be generally so. Table 8.10 relates only to people making their first visit to each museum—people whose visits, as we have just seen, tend to be comparatively wide-ranging. It does somewhat understate the extent of this group's visits since it is based on the replies only of those people who were able to give complete routes. One of the reasons that some visitors were unable to provide full information was that their visits had been so extensive that they could not remember all the galleries they had walked through. Nevertheless the message of Table 8.10 is clear: the majority of first-time visitors to each London museum walked through less than half the museum's 40 or more galleries. At best they would have glanced into the remaining galleries from a distance.

Table 8.11 lists the Victoria and Albert Museum's galleries in decreasing order of theg proportions of all the museum's visitors who stopped to look at each. It is possible (see Chapter 9, Section (a)) that attendance at some of the museum's galleries is understated. However, even with this qualification, Table 8.11's figures reveal a definite pattern. This can be seen by looking down the table and noting the position of each gallery on the plan of the museum (see pages 2–3). The rule is that the nearer the entrance a gallery is situated the more visitors it receives. The Ironwork gallery, the fifteenth in the list, is the first to be located in the half of the plan devoted to the upper floors. The Silver gallery, the twenty second in the list, is the first to be situated at the back of the museum. If we include the museum's shop, seven of the ten most viewed galleries are on the ground floor at the front of the museum. The other three—the galleries devoted to Oil Paintings, Constable and Continental Eighteenth Century Art— are located in rooms that can be reached via short staircases leading down directly from the entrance hall.

The pattern at the Science Museum is rather more complex. Distance from the entrance appears to play some part in the sense that galleries on the ground floor are generally looked at by more visitors than galleries on the higher floors. On the other hand horizontal distance from the entrance does not seem to matter. Galleries at the back of the museum are as well attended as, or indeed better attended than, galleries at the front of the museum.

One reason for this is the attractive power of the galleries devoted to the few 'star' subjects that we noted in Chapter 3. The Exploration gallery, with its emphasis on space travel, the galleries devoted to road and rail transport, as well as the Aeronautics gallery on the third floor and the Computing Then and Now gallery on the second floor, are all near the top of the list in Table 8.12. The Children's gallery was rather less frequently visited by members of the sample but still occupies fourteenth place, quite well up the list. (It is worth recalling that children aged ten or less were not included in the quotas of people to be interviewed.) All these permanent 'star' galleries are situated towards the back of the museum.

Leaving aside the 'star' galleries, 15 of the remaining 34 permanent galleries are towards the back of the museum, that is their position on the plan is to the left of the major staircase that leads up from the ground floor opposite the Information Office. The other 19 permanent galleries are in the front part of the museum. Even with the 'star' galleries left out of account there is no tendency for galleries toward the front of the museum to be better attended than galleries at the back. In fact ten of the 15 galleries at the back of the museum were visited (ie stopped to look at) by 11 per cent or more of the people we interviewed, compared with only seven of the 19 galleries at the front of the museum.

Table 8.10   The number of galleries first-time visitors to the two London museums walked through or stopped to look at

Number of galleries* walked through or stopped to look at	Victoria and Albert Museum	Science Museum
	%	%
1–5	13	10
6–10	27	22
11–15	29	28
16–20	21	19
21–25	7	10
26 or more	3	11
Base†	363	308

* The entrance hall, museum shop and restaurant are counted as galleries.

† Note that this table excludes visitors who could not record an apparently complete route.

Source of table: the quota samples of museum leavers.

A corollary of this is that signposts and other directional information are likely to be important to visitors who want to ensure that their limited coverage of the museum takes in the galleries dealing with topics that interest them. With no directional information visitors would be likely to miss much of interest altogether. We shall return to the importance of signposting in subsequent chapters.

d. Choosing which galleries to look at
We have already noted in Chapter 3 one determinant of visitors' choice of which galleries to look at—namely the fact that many visitors set out for the museums with the intention of seeing the galleries devoted to specific subjects. Tables 8.11 and 8.12 based on the quota sample interviews with museum leavers, illustrate the importance of another determining factor—the physical geography of the museums themselves.

Table 8.11 The proportion of visitors who stopped to look at each gallery in the Victoria and Albert Museum

Gallery and floor*	Percentage of the museum's visitors who stopped to look at something in the gallery
	%
Shop (G)	55
Oil Painting (LG)	27
Constable (LG)	26
Early Medieval Art (G)	25
Gothic Art: Italy, England, France, Germany (G)	25
Continental Eighteenth Century Art (LG)	23
British Sculpture (G)	23
Art of China and Japan (G)	22
Art of Islam (G)	21
Temporary Exhibition gallery A (G)	20
British Art: Tudor and Early Stuart (UG)	19
Carpets (G)	19
Continental Sculpture (G)	17
British Art 1650–1750 (UG)	17
Ironwork (1)	16
Raphael Cartoons (G)	16
Indian Sculpture (G)	16
Woodwork and Architectural Study Collection (G)	15
High Renaissance (G)	14
Continental Seventeenth Century Art (LG)	14
Indian Art (G)	13
Silver (1)	13
Prints and Drawings (1)	13
Stained Glass (1)	13
Spanish Gothic (G)	12
Embroidery (1)	12
Casts (passage through centre) (G)	11
Renaissance Italy (G)	10
Textile Study Rooms (1)	10
Renaissance Northern Europe (G)	9
Metal Work (1)	8
Tapestries (1)	8
German Stained Glass (1)	8
Mixed Indian and South East Asian (G)	7
Medieval Tapestries (G)	7
Temporary Exhibition gallery B (G)	7
Armour (1)	7
Victorian Art (U1)	7
Musical Instruments† (1)	7
Art of the Book (1)	6
Glass Vessels (1)	5
English and European Ceramics (2)	5
British Art 1750–1820 (U1)	5
Carvings, Sculpture and Bronzes (UG)	4
French Ceramics (U1)	4
Morris, Gamble and Poynter Rooms (G)	4
Garden (G)	4
Far Eastern Ceramics (2)	4
Chinese Stone Carvings (U1)	3
Twentieth Century Study Collection (1)	3
Watercolours (1)	3
Ancient and Near Eastern Pottery† (2)	2
Base	1,001

* LG = Lower ground floor   1 = First floor
   G = Ground floor   U1 = Upper first floor
   UG = Upper ground floor   2 = Second floor

† *Attendance in the Musical Instruments gallery and in the Ancient and Near Eastern Pottery gallery was lower than it otherwise would have been since they were closed for part of the survey year.*

Source of table: the quota sample of museum leavers.

Table 8.12 The proportion of visitors who stopped to look at each gallery in the Science Museum

Gallery and floor*	Percentage of the museum's visitors who stopped to look at something in the gallery
	%
Exploration (G)	70
Rail Transport and Trams (G)	53
Development of Motive Power (G)	51
Aeronautics (3)	42
Road Transport (G)	41
Fire Fighting Appliances (G)	36
Computing Then and Now (2)	33
Temporary Exhibition gallery (1)	32
Bicycles and Motor Cycles (G)	31
Communications (3)	27
Shop (G)	27
Temporary Exhibition gallery (G)	24
Shipping (main floor) (2)	24
Children's gallery (LG)	22
Agriculture (1)	22
Navigation (2)	20
Astronomy (1)	18
Photography and Cinematography (3)	18
Domestic Appliances (LG)	18
Gas gallery (1)	17
Iron and Steel (1)	16
Time Measurement (1)	14
Glass Technology (1)	14
Optics (3)	13
Space Exploration and Rocketry (3)	11
Hand and Machine Tools (1)	11
Pure Chemistry (2)	11
Meteorology (1)	11
Printing and Paper (2)	10
Textile Machinery and Sewing machines (1)	10
Firemaking (LG)	10
Electric Power (G)	10
Shipping (raised floor) (2)	9
Locks and Fastenings (LG)	8
Atomic Physics and Nuclear Power (2)	8
Heat and Thermal Instruments (3)	8
Industrial Chemistry (2)	8
Early Physics (3)	7
Mapmaking and Surveying (1)	6
Lighting (2)	5
Geophysics (3)	5
Roads, Bridges and Tunnels (G, raised)	5
Talking Machines (3)	4
Radio Room (3)	4
Temporary Exhibition gallery (3)	3
Weights and Measures (2)	2
Raised Space for Temp Exhibitions (G)	1
Base	1,044

* LG = Lower ground floor   2 = Second floor
   G = Ground floor   3 = Third floor
   1 = First floor

Source of table: the quota samples of museum leavers.

Why is it that even non-'star' galleries at the back of the Science Museum are as well attended as galleries at the front? Part of the explanation may be that the architecture of each floor is simple enough for visitors to feel free to wander all over it once they have made the effort to reach the floor in the first place. Another part of the explanation is a knock-on effect of the attractive power of the star galleries. On their way to and from the 'star attractions' visitors are likely to pause to look at neighbouring galleries whose contents catch their attention. Since the 'star' galleries are at the back of the museum this knock-on effect may raise the attendance at other galleries in that part of the museum above what it otherwise would be.

Table 8.13 shows this process at work. The table focuses on the Aeronautics gallery and the galleries close to it on the same floor and the floor below. Comparing the attendance of visitors who had arrived at the museum specifically intending to see the Aeronautics gallery with that of other people, it shows that people who set out to visit the Aeronautics gallery were much more likely than other people to have stopped to look at the neighbouring galleries. It seems that many of the visitors attracted by the Aeronautics gallery also stopped to look at the galleries they passed through on their way to and from it.

Table 8.13 **Proportions of the Science Museum's visitors who stopped to look at the Aeronautics gallery and neighbouring galleries by whether or not they had planned to visit the Aeronautics gallery when setting out for the museum**

Gallery and floor*	Visitors who planned to visit Aeronautics gallery	Other visitors
Aeronautics (3)	98%	39%
Space Exploration and Rocketry (3)	19%	11%
Communications (3)	48%	26%
Shipping (raised floor) (2)	16%	8%
Shipping (main floor) (2)	43%	23%
Navigation (2)	24%	20%
Base	63	981

*See note on Table 8.12.
Source of table: the quota samples of museum leavers.

Table 8.14 **Proportion of visitors who had come to the Victoria and Albert or Science Museum to see a special temporary exhibition by whether they had visited the museum before**

	First visit (a)	Been before (b)	Weighted bases	
			(a)	(b)
Victoria and Albert Museum	13%	38%	470	530
Science Museum	5%	14%	460	540

Source of table: weighted results from the count-based samples of museum leavers.

e. Special temporary exhibitions

Special temporary exhibitions play a considerable part in attracting visitors to the Victoria and Albert Museum and, though to a lesser extent, to the Science Museum. Table 3.1 in Chapter 3 indicated that 26 per cent of the Victoria and Albert Museum's visitors, and ten per cent of people visiting the Science Museum, cited a special exhibition as a main reason for their visit. Because special exhibitions have such a major role it is important to consider the people they attract. In particular do they, as the above figures may suggest, attract many people to the museums who would not have visited them otherwise?

Table 8.14 confirms that special exhibitions do attract some first-time visitors to the two museums. Thirteen per cent of people making their first visit to the Victoria and Albert Museum and five per cent of the Science Museum's new visitors mentioned a special exhibition as a main reason for their visit.

However the main role of special exhibitions seems to be persuading people who have already been to the museum concerned to keep visiting. Thirty-eight per cent of people revisiting the Victoria and Albert Museum had come to see a special exhibition—approximately three times the proportion of new visitors who gave this as a reason. In the Science Museum too, return visitors were roughly three times as likely as first-time visitors to have come to see a special exhibition.

f. Reasons for the seasonal variations in visiting at the National Railway Museum

The National Railway Museum's own attendance figures show that it receives far more visitors during the summer than it does in the winter—prompting interest in why people choose particular times of the year for their visits. Table 8.15 sets out the replies visitors gave to the question:

"Is there any particular reason why you chose this time of year for your visit?"

Forty-five per cent of the visitors gave no reason for visiting at the time of year they did. However the answers given by the remaining 55 per cent suggest that two factors account between them for much of the seasonal variation in attendance.

Table 8.15 **Reasons given by visitors to the National Railway Museum for choosing to visit at the time of year they did by whether they were visiting alone or in a group**

Reason given	Visiting:					All visitors
	Alone	With friends but not members of family	With members of family	With a school party	With another type of organised party	
	%	%	%	%	%	%
Holiday	23	22	45	0	12	33
Date of institutional visit	0	0	1	25	32	6
Visiting York	12	3	6	0	5	5
Less crowded	3	2	4	2	0	3
No reason given	45	57	38	68	45	45
Weighted base	50	110	640	140	60	1,000

Note that visitors could give more than one answer to the question on which this table is based. However percentages add to less than 100% because reasons mentioned by less than 3% of the weighted sample have been omitted from this table.
Source of table: weighted results from the count-based samples of museum leavers.

People visiting with organised parties tended, if they gave a reason at all, to say that this was the day on which their school—or some other institution to which they belonged—was visiting the museum. Essentially the date of the visit has been fixed by the institution itself and was not up to them.

By far the most important factor for the museum's other visitors was the timing of their holidays. Holidays played a particularly important part in setting the dates of family visits. Forty-five per cent of family visitors mentioned that they, or some other member of their family, were on holiday.

The sources of this information were visitors interviewed in the count-based sample of museum leavers. As we only carried out interviews of this type on 16 of the days in the survey year it is not really possible to break the figures down by time of year. However it is clear that the answers recorded in Table 8.15 would explain why many visitors choose the summer months—the time of the main family holiday—for their visits.

9 Patterns of interest—visitors' reactions to all the widely visited galleries

a. General

Chapters 4 to 7 studied visitors' reactions to a few galleries in depth. In the National Railway Museum the 'galleries' studied did in fact include nearly all the museum's exhibition space. The situation in the two London museums was quite different—only four of the 40 or more permanent galleries in each museum were selected for detailed study. It is now time to widen the focus in order to say something about visitors' reactions to the remaining galleries. Since the National Railway Museum was fairly completely covered by the 'gallery' samples it will not be discussed in this chapter.

Chapter 8 explained how members of the quota samples of people interviewed when they left each museum were shown a plan of the whole museum and asked to identify the galleries in which they had 'stopped and looked' at something. They were then asked to rate each gallery according to whether they found it 'very interesting', 'fairly interesting' or 'not really interesting'. Tables 9.1 and 9.2 show, for the Victoria and Albert Museum and the Science Museum respectively, the percentage of each gallery's visitors who rated it 'very interesting'. The tables are restricted, in the interests of the reliability of the results, to galleries for which we obtained interest level ratings from over a 100 visitors. This means that interest ratings are not given for the galleries listed towards the bottom of Tables 8.11 and 8.12. Because the quota samples did not include children aged ten or under, the interest levels given in Tables 9.1 and 9.2 do not reflect their views.

The function of these tables is two-fold. First they identify the galleries which appeal most to the people who visit them and distinguish these from the galleries which have been less successful in appealing to the majority of their visitors. Secondly the tables provide an opportunity to test some of the ideas developed in Chapters 4 to 7 about what it is that makes particular galleries interesting. This can be done by considering the galleries higher up the list and asking whether they have the qualities that the earlier chapters have suggested would help to make galleries interesting, and by asking whether the galleries lower down the list are handicapped by any features that have been identified as going with a low level of interest.

Deciding what are the relevant features of each gallery is purely a matter of judgement since we were not able to ask the quota samples of museum leavers what they considered the salient characteristics of each gallery.

Table 9.1 Percentages of visitors rating each Victoria and Albert Museum gallery 'very interesting'

Gallery and floor*	Percentage rating gallery 'very interesting'	Base
Constable (LG)	75%	138
Continental 17th Century (LG)	73%	133
British Art: Tudor and Early Stuart (UG)	69%	191
Continental 18th Century (LG)	68%	232
British Art 1650–1750 (UG)	65%	167
Art of China and Japan (G)	63%	222
Oil Paintings (LG)	62%	266
Woodwork and Architectural Study Collection (G)	62%	148
Silver (1)	62%	133
Stained Glass (1)	58%	126
Indian Art (G)	55%	134
Raphael Cartoons (G)	54%	159
Embroidery (1)	50%	118
Art of Islam (G)	46%	214
Ironwork (1)	45%	165
Early Medieval Art (G)	43%	255
Continental Sculpture (G)	42%	172
Carpets (G)	41%	188
British Sculpture (G)	41%	230
Gothic Art: Italy, England, France, Germany	40%	247
Spanish Gothic (G)	39%	121
Indian Sculpture (G)	38%	157
High Renaissance (G)	31%	138

See note on Table 8.11.

This table covers the permanent galleries which a hundred or more members of the sample stopped to look at. The Cast Court is excluded from the table as only a narrow central corridor was open during the survey year.

Source of table: the quota samples of musem leavers.

In practice this means that it is probably best for someone who knows each gallery well to identify factors which might affect interest in that gallery specifically. However it is possible to go some way towards explaining the distribution of interest level ratings by considering fairly superficial characteristics of the different galleries. The following two sections of the present chapter will attempt to do this.

However before using Tables 9.1 and 9.2 at all it is necessary to have some confidence in the validity of the interest level ratings given by people as they leave the museum. We need to be sure that they can remember their feelings about the individual galleries clearly enough to give valid ratings. Fortunately we can investigate the effect of asking for interest ratings as the visitors leave the museum, rather than as they leave the gallery concerned, by comparing the ratings that the museum leavers gave the galleries which were studied in depth to the ratings given by the samples of visitors who were questioned as they left each gallery.

Table 9.2 Percentages of visitors rating each Science Museum gallery 'very interesting'

Gallery and floor*	Percentage rating gallery 'very interesting'	Base
Exploration (G)	67%	736
Children's gallery (LG)	64%	230
Aeronautics (3)	64%	443
Road Transport (G)	61%	427
Rail Transport and Trams (G)	58%	553
Domestic Appliances (LG)	58%	189
Bicycles and Motor Cyles (G)	56%	319
Computing Then and Now (2)	56%	345
Photography and Cinematography (3)	54%	190
Shipping (2)	54%	250
Space Exploration and Rocketry (3)	51%	120
Communications (Telegraph, Telephone, Radio, TV, Radar) (3)	46%	286
Optics (3)	43%	139
Printing and Paper (2)	42%	109
Time Measurement (1)	42%	149
Electric Power (G)	42%	101
Astronomy (1)	41%	193
Glass Technology (1)	41%	141
Fire Fighting Appliances (G)	41%	374
Navigation (2)	40%	210
Agriculture (1)	38%	226
Pure Chemistry (2)	37%	113
Development of Motive Power (G)	32%	537
Firemaking (LG)	31%	104
Meteorology (1)	29%	112
Gas gallery (1)	29%	176
Hand and Machine Tools (1)	28%	115
Textile Machinery and Sewing Machines (1)	27%	107
Iron and Steel (1)	26%	171

* See note on Table 8.12.
This table covers the permanent galleries which a hundred or more members of the sample stopped to look at.
Source of table: the quota samples of museum leavers

According to Table 9.2 the Exploration, Aeronautics, Printing and Paper and Time Measurement galleries were found very interesting by respectively 67 per cent, 64 per cent, 42 per cent, and 42 per cent of their visitors. Assessments obtained from people as they left these galleries—reported already in Table 4.4—were 64 per cent, 61 per cent, 32 per cent and 33 per cent respectively. The fact that the two sets of figures display such a similar pattern suggests that visitors leaving the Science Museum had little difficulty recalling their opinions of specific galleries. It is worth noting that the four galleries selected for special study did not include any with really low ratings for interest. The Exploration gallery occupies first place in the list of Table 9.2 and the Aeronautics gallery is third. The Printing and Paper and Time Measurement galleries come about half way down the list.

Table 9.1 gives the following percentages of visitors rating as 'very interesting' the four Victoria and Albert Museum galleries that were studied in depth: Continental Seventeenth Century Art 73 per cent, British Art: Tudor and Early Stuart (which included the rooms referred to as the Tudor Art gallery in Chapters 4 to 7) 69 per cent, Art of China and Japan 63 per cent and British Sculpture 41 per cent. The corresponding percentages derived from the answers of people inter-

viewed as they left the galleries concerned were given in Table 4.1 as, respectively, 50 per cent, 44 per cent, 52 per cent and 37 per cent.

The figures are not as reassuring as those for the Science Museum. Although the relative ranking of the galleries obtained from answers at the exit places the galleries in much the same order as that obtained from interviewing people on the spot—Continental Seventeenth Century Art being more popular than Tudor Art which is more popular than British Sculpture, only the relative position of the Art of China and Japan has changed—some of the galleries receive much higher ratings from the sample interviewed at the museum's exit than they do from those interviewed on the spot. This difference is over 20 per cent for the Tudor and Continental Seventeenth Century Art galleries, but only 11 per cent for the Art of China and Japan gallery and four per cent for the British Sculpture gallery.

We cannot be sure of the reasons for this apparent over-estimation of galleries' interest levels by people leaving the museum but an explanation that could account for the general over-estimation and for its apparent tendency to be more pronounced in some galleries than in others would run as follows.

We noted in Chapter 4 that many visitors interviewed leaving galleries where they had 'stopped to look at something' had in fact spent only a very brief time there. They tended, as we saw in Chapter 5, to be people for whom the gallery had little appeal who had stopped just long enough to decide that they did not find it particularly interesting. If visitors like these had forgotten these brief reconnoitres by the time they came to leave the museum, their low interest level ratings for the gallery would be excluded from the figures collected from people leaving the museum—pushing up the gallery's apparent interest rating. It might well be easier to forget such brief reconnoitres in the Victoria and Albert Museum where many galleries (as was noted in Chapter 4) shade off into one another than in the Science Museum where they are clearly marked off by changes in subject matter. The comparatively complex structure of the Victoria and Albert Museum may also have made it hard to locate some less vividly remembered galleries on the plan. It would not be surprising if short visits to the Art of China and Japan gallery with its distinctive subject matter and to the British Sculpture gallery which is very near the museum's entrance were better remembered than those to the other two galleries.

The strong possibility that the positive bias in the interest levels obtained from the sample of people leaving the Victoria and Albert Museum affects the ratings given to different galleries to varying extents means that the figures in Table 9.1 have to be treated with considerable caution. It would be unwise to place any emphasis on differences of a few per cent between the interest level ratings given to two galleries in different parts of the Victoria and Albert Museum.

b. Galleries found more and less interesting at the Victoria and Albert Museum

Despite the strictures of the last section there are a number of conclusions which can safely be drawn from Table 9.1. Among the galleries receiving higher ratings for interest are the Constable gallery on the lower ground floor, and the gallery leading to it—which displays oil paintings by a variety of artists.

Other galleries rated highly by their visitors are the rooms devoted to Continental Eighteenth Century Art which can be entered by steps leading down from the left of the entrance hall and, leading on from them, the Continental Seventeenth Century Art gallery which is one of the galleries we have studied in the previous four chapters. Both galleries contain examples of many different art forms though both place a fairly strong emphasis on furniture. The eighteenth century gallery makes great use of period room settings, though the seventeenth century gallery uses them less. The atmosphere of the eighteenth century gallery, like that of the seventeenth century gallery, is carefully controlled for temperature and humidity. Both galleries present a rather sumptuous appearance.

The two upper ground floor galleries devoted to British Art in Tudor and early Stuart times and from 1650 to 1750 can definitely be included among those which are relatively popular with their visitors. Like the two continental art galleries immediately beneath them they contain many different art forms with some emphasis on domestic furnishings. These are mostly displayed in the form of room settings—the objects being arranged to give the idea of very prosperous domestic interiors of the historical periods concerned.

There are four galleries which form in effect a single long corridor running right across the ground floor parallel to the front of the museum and dividing the less visited rear part of the museum from the rest. Three of these galleries—those covering the High Renaissance, Gothic Art in Italy, England, France and Germany, and Spanish Gothic—are devoted to works of sculpture in the styles which their names suggest. The fourth, leading on from the Spanish Gothic gallery, displays carpets. All four galleries received comparatively low interest ratings from their visitors. So did four other ground floor galleries devoted largely to sculpture or carving—The British and Continental Sculpture galleries, the Indian Sculpture gallery, and the Early Medieval Art Gallery at the centre of the ground floor behind the museum's shop. The only exception to the low interest ratings which sculpture galleries received from their visitors was the Woodwork and Architectural Study collection which contained a good deal of carving.

These ratings do tend to confirm some of the conclusions that were drawn in the last few chapters from the detailed studies of visitors' reactions to selected galleries. We saw in Chapter 5 that visitors with a prior interest in the subject of a gallery were much more likely than other visitors to feel that their visit to the gallery had been very interesting. One would therefore expect that galleries devoted to subjects which particularly drew visitors to the museum would receive higher interest ratings than galleries whose subject matter had less initial appeal to most of the museum's current visiting public. It was noted in Chapter 3 that paintings and furniture both played a greater part than sculpture in drawing visitors to the Victoria and Albert Museum. It is therefore no surprise that galleries displaying furniture and paintings tend to receive higher ratings than galleries devoted to sculpture.

Chapter 6 suggested that attractive presentation could add considerably to the interest of a gallery and that among the features that led to a gallery being judged attractive were a sense of spaciousness, good lighting and, where they applied, period room settings. Some or all of these features do seem to be applicable to each of the galleries that received particularly high interest ratings. By contrast the four galleries which form a corridor and the Indian Sculpture gallery are rather dark, and the Gothic Art gallery might also strike some visitors as cluttered.

The ratings obtained from people leaving the Victoria and Albert Museum are therefore consistent both with the previous conclusion that prior interest in the subject of a gallery is very important and with the conclusions drawn about the types of presentation that are likely to be most effective. However since the galleries which one would expect on the basis of their subject matter to receive high interest ratings tended also to be presented in ways that seemed likely to be effective, the ratings given by people leaving the museum cannot be used to estimate the relative importance of the contributions of subject matter on the one hand and presentation on the other.

c. Galleries found more and less interesting at the Science Museum

In Chapter 3 we saw that there were five subject areas or galleries which visitors to the Science Museum were particularly likely to want to see. These were space exploration, road and rail transport, aircraft, computers, and the Children's gallery. Eight of the 11 permanent Science Museum galleries most highly rated by their visitors for interest dealt with these subjects.

Space exploration was a major topic of the Exploration gallery which comes top of the list in Table 9.2 as well as of the small Space Exploration and Rocketry gallery on the third floor which ranks eleventh. The Children's gallery in the basement and the Aeronautics gallery on the third floor were rated very interesting by almost as high a proportion of their visitors as the Exploration gallery. (It is noteworthy that the Children's gallery achieved this high rating even though children aged ten or under were not included in the sample that was asked about interest levels.) The three ground floor galleries devoted to Bicycles and Motor Cycles, Road Transport (mainly cars) and to Rail Transport and

Trams were each considered very interesting by between 56 and 61 per cent of their visitors. Computing Then and Now, on the second floor, was thought very interesting by 56 per cent of its visitors.

The subjects of three of the galleries out of the 11 most highly rated for interest were not ones which had played a major role in attracting visitors to the museum. The galleries concerned were Domestic Appliances in the basement which was found very interesting by 58 per cent of its visitors as well as the main floor of the Shipping gallery on the second floor of the museum which was highly rated by 54 per cent of its visitors as was Photography and Cinematography on the third floor.

The Communications gallery on the third floor which deals with the development of electronic communications and the gallery on the second floor which deals with the closely related subject of aids to navigation were among the galleries receiving 'very interesting' ratings from between 40 and 46 per cent of their visitors. Other galleries with interest ratings in this range were the Optics gallery on the third floor; the Printing and Paper gallery on the second floor; the Astronomy, Time Measurement and Glass Technology galleries on the first floor; and, on the ground floor, the Electric Power gallery and the gallery which shows fire engines and firemen's protective gear.

Six of the last 9 galleries listed in Table 9.2 are on the first floor. They include the Agriculture gallery 38 per cent of whose visitors rated it very interesting, as well as five galleries each of which was thought very interesting by less than a third of the people who stopped to look at it. These were the galleries devoted to Meteorology, Gas, Hand and Machine Tools, Textile Machinery and Sewing Machines, and Iron and Steel.

The three remaining galleries are the Pure Chemistry gallery on the second floor, the Development of Motive Power gallery on the ground floor down some steps from the museum's main entrance, and the Firemaking gallery in the basement—each of which was found very interesting by between 30 and 40 per cent of its visitors.

The fact that the galleries receiving the highest ratings for interest were nearly all devoted to popular subjects illustrates again the important part that the interests visitors bring with them play in determining how they react to the galleries—confirming the result obtained in Chapter 5. The next question to ask is whether the table provides any confirmation of the conclusions about presentation that were reached in Chapter 6. The fact that the Exploration gallery receives a higher rating than the Space Exploration and Rocketry gallery on the third floor—which also deals with the most popular aspect of the Exploration gallery's subject matter—suggests that presentation may have some importance.

The highly rated Domestic Appliances gallery might almost have been chosen to illustrate the main points made in Chapter 6. The subject matter is familiar to everyone and this combined with comparatively straightforward labels and diagrams probably meant that few of the gallery's visitors thought it was meant for people who knew more than they did themselves. The labels are made easier to read by the use of large print. The gallery is spacious and even lighting makes it possible to see all the exhibits without difficulty. At one end of the gallery are some full-scale room settings including a mock-up of a Victorian kitchen showing a maid using the equipment it contains. A loudspeaker commentary details what her duties would have been. These room settings could be described, in the terms used in Chapter 6, as a special effect giving an added sense of reality to the exhibits.

The Photography and Cinematography gallery—another of the galleries which achieved a high interest rating despite not having been a major reason for visiting the museum—is also laid out with an eye to effect. It consists of two darkened circular rooms each with cases round the sides set at right angles to the walls. At the centre of each room is a major set-piece exhibit. Each case contains equipment relating to some clearly marked specific aspect of photography or cinematography. There are a few television and slide recordings accompanied by loudspeaker commentaries.

The Shipping gallery which received the same proportion of 'very interesting' ratings as the Photography and Cinematography gallery has a rather more conventional lay-out. The gallery contains ship models—often displayed in dioramas—as well as various pieces of nautical equipment and, hanging above the visitors' heads, a fast rowing boat and one or two other light craft. Traditional as the presentation is, it does have the attributes which Chapter 6 identified as important. It is well lit and fairly spacious. The written information is easy to understand and the dioramas combine with a number of the major exhibits to give the gallery a distinctly nautical atmosphere.

However, although some of the more highly rated galleries do seem to be well presented, not all the galleries which, applying the criteria of Chapter 6, seem well presented succeed in obtaining high interest ratings from their visitors. The Agriculture gallery which (except for being rather more cramped) uses much the same kind of presentation as the Shipping gallery was rated very interesting by only 38 per cent of its visitors. It seems possible that agriculture is simply less interesting than shipping to many of the museum's visitors.

An even clearer illustration of the limitations to what can be achieved by attractive presentation is provided by the very last gallery listed in Table 9.2—the Iron and Steel gallery rated 'very interesting' by only 26 per cent of its visitors. From the point of view of attractiveness the gallery seems to do all that other galleries do to achieve higher ratings. Special effects are provided by full-scale models of the opening of a furnace and of a red hot ingot being lifted by machinery. The gallery is adequately lit and not cramped. The explanations are in large clear print. However despite all this the gallery's appeal is very limited. Again one plausible explanation is that the gallery's subject has little intrinsic interest for its visitors. Another possibility is that, though the explanatory material is clearly printed, many visitors find the text itself—which uses a good deal of technical language—rather over their heads. As Chapter 6 showed, this would be likely to limit their interest in the gallery considerably.

The findings of this section seem to be broadly consistent with those of Chapters 5 and 6. They suggest that the intrinsic appeal of each gallery's subject matter plays a crucial role in determining visitors' interest in the gallery. While it seems likely that the points of presentation identified in Chapter 6 can enhance a gallery's appeal it should not be assumed that attractive presentation is enough in itself to interest visitors in a subject they think dull or difficult to understand.

10 Visitors' use of facilities and their views on practical aspects of how the museums are run

The preceding chapters have dealt with the ways in which visitors set about looking at the museums' exhibits and have focused on a series of related issues concerning viewing patterns, interest levels and methods of presentation. This chapter will try to throw light on a rather varied collection of practical issues which may also make important contributions to people's enjoyment of their visits. In order to avoid unnecessary repetition the information is arranged by topic rather than by museum.

a. Helping visitors find their way around

Members of the quota samples of museum leavers were asked whether they thought that improvements needed to be made in the signs provided to help people find their way round the museum. As Table 10.1 shows, half of the Victoria and Albert Museum sample and somewhat less than half the Science Museum sample did think the signs needed improvement. Only 12 per cent of the National Railway Museum sample thought that improvements were needed. At all three museums the proportion seeing a need for improvements was as high amongst those who had been to the museum before as it was amongst first-time visitors.

As the figures in Table 10.2 indicate, the great majority of those wanting improvements thought that the museums should put up more signs or display them more prominently. However at the Victoria and Albert Museum and at the Science Museum there were also substantial groups (18 per cent and 23 per cent respectively of those wanting the signs improved) who felt that the existing signs were confusing and should be made clearer.

In addition a number of visitors made more specific suggestions, the most common at the two South Kensington museums being that more large museum plans should be displayed. Several visitors to the National Railway Museum would have liked signs indicating which direction to follow. This may perhaps be because in many parts of the Main Hall the nearby locomotives or rolling stock tower over the visitor, blocking his view of the rest of the hall's contents.

Table 10.2 Ways in which the signs were felt to need improvement: comments of sample members who felt improvements were needed

Improvements felt to be necessary	Victoria and Albert Museum	Science Museum	National Railway Museum
	%	%	%
More, or more prominent, signs	79	71	60
Clearer signs	18	23	7
(More) museum plans on walls, etc	9	9	8
Indication of direction to follow	4	3	29
Colour-coding	4	2	0
Other	17	13	12
Base	516	416	73

Percentages add to over 100% as people could give more than one answer.
Source of table: the quota samples of museum leavers.

However it would be wrong to attach great importance to the fact that 29 per cent of those members of the National Railway Museum sample who felt the signs needed improving suggested this particular improvement, since they only represent four per cent of the whole quota sample of people leaving the National Railway Museum.

Table 10.3 lends weight to the contention that directional information is not always displayed as prominently as it might be in the two South Kensington museums. Forty one per cent of the sample at the Victoria and Albert Museum had not realised at the start of their visit that a plan of the museum was displayed in the entrance hall—a figure which rose to 48 per cent among those visiting for the first time. The plan at the Victoria and Albert Museum was placed at the back of the entrance hall to the right of the entrance to the shop and the approach to it was sometimes partly blocked by seats. However the plan was made fairly noticeable by using different colours, being placed at eye level and by being more or less straight ahead of visitors entering the museum. The plan at the Science Museum was black and white, near floor level and situated at the back of the entrance hall some way off to the left-hand side. The proportion of visitors

Table 10.1 Percentage of sample who felt improvements were needed in the signs provided to help people find their way round by whether the visitor had been to the museum before

| | Victoria and Albert Museum (a) | Science Museum (b) | National Railway Museum (c) | Bases | | |
				(a)	(b)	(c)
First-time visitors	48%	36%	12%	469	378	386
Others	55%	42%	12%	534	666	210
All visitors	52%	40%	12%	1,003	1,045	596

Source of table: the quota samples of museum leavers.

Table 10.3 Awareness and take-up of museum plans

Visitors, who at the start of their visit:	Victoria and Albert Museum			Science Museum		
	First-time visitors	Others	All visitors	First-time visitors	Others	All visitors
Were aware there was a plan in the entrance hall	52%	65%	59%	25%	32%	30%
Realised they could buy plans	43%	53%	48%	23%	31%	28%
Did buy plans*	22%	12%	17%	7%	5%	6%
Base	*469*	*534*	*1,003*	*378*	*666*	*1,045*

* *Does not include visitors who bought guide books.*
Source of table: the quota samples of museum leavers.

Table 10.4 Awareness of lifts and escalators to upper floors

	Victoria and Albert Museum			Science Museum		
	First-time visitors	Others	All visitors	First-time visitors	Others	All visitors
Visitors who realised at the start of their visit that there were lifts (and escalators at the Science Museum)	23%	30%	27%	42%	58%	52%
Base	*469*	*534*	*1,003*	*378*	*666*	*1,045*

Source of table: the quota samples of museum leavers.

whose attention it attracted was correspondingly low—only 25 per cent of the first-time visitors in the sample having been aware of it. Still fewer visitors to the two South Kensington museums—only 43 per cent of the first-time visitors in the Victoria and Albert Museum sample and 23 per cent at the Science Museum—had realised that they could buy plans of the museum to take round with them. Half of those first-time visitors to the Victoria and Albert Museum who had realised that the plans were available had in fact bought one. As many visitors who did not buy a plan themselves will have had a companion who did, this suggests that well over half the Victoria and Albert Museum's first-time visitors made use of the plans when they were aware of their existence. The seven per cent of first-time visitors in the Science Museum sample who bought plans represent over a quarter of those who knew about them, so that at the Science Museum too considerably over a quarter of first-time visitors who realised that there were plans to be had must have used them. If the availability of these plans was more prominently announced at both museums, it seems likely that the take-up would be greatly increased.

Table 10.4 shows that many visitors to the two South Kensington museums had not realised at the start of their visits that they could reach the upper floors without climbing the stairs. Their failure to realise this may help explain the comparative scarcity of visitors on the higher floors.

The figures in Table 10.1 show that people who have visited the museum before are at least as likely as first-time visitors to think that the museum's signposting system needs improvement. Table 10.5 shows that they are also just as likely as first-time visitors to ask a

Table 10.5 Percentage of sample who asked a member of staff for help in finding their way around by whether the visitor had been to the museum before

	Victoria and Albert Museum (a)	Science Museum (b)	National Railway Museum (c)	Bases		
				(a)	(b)	(c)
First-time visitors	35%	22%	4%	469	378	386
Others	38%	26%	1%	534	666	210
All visitors	36%	24%	3%	1,003	1,045	596

Source of table: the quota samples of museum leavers.

Table 10.6 Awareness of museums' information services

Visitors who were aware before the start of their current visit that the museum:	Victoria and Albert Museum			Science Museum		
	First-time visitors	Others	All visitors	First-time visitors	Others	All visitors
Put on public lectures	28%	77%	54%	24%	62%	48%
Had a library open to the public	16%	41%	29%	10%	23%	18%
Had expert staff who gave opinions on objects brought by public	21%	59%	41%	16%	34%	28%
Base	*469*	*534*	*1,003*	*378*	*666*	*1,045*

Source of table: the quota samples of museum leavers.

member of staff for directions. Thirty-six per cent of the Victoria and Albert Museum sample and 24 per cent of the sample at the Science Museum had asked a member of staff for help in finding their way round. It seems that the museum staff play an important part in filling the gaps in the systems of directional signs.

b. Information services

The figures in Table 10.6 show how far members of the quota samples of people leaving the two South Kensington museums were aware of certain information services provided by each museum. Visitors were asked whether they had known of each service before the start of the present visit. First-time visitors were much less knowledgeable about all these services than were members of the samples who had visited the museum before—which could suggest either that people mainly find out about these services by seeing notices about them in the museums themselves (which first-time visitors could not have done before their present visit) or that the people who are sufficiently motivated to find out about these services tend to be found amongst those sufficiently interested in the museum to pay it one or more return visits.

Concentrating on sample members who had visited the museums before, we see that the existence of public lectures at each museum was comparatively well known—77 per cent of this group having been aware of them at the Victoria and Albert Museum and 62 per cent at the Science Museum. At the Victoria and Albert Museum 59 per cent had been aware that its staff would provide expert opinions on objects which the public brought them. The other services were less well known, comparable figures being 41 per cent and 23 per cent respectively having known that the libraries at the Victoria and Albert Museum and the Science Museum were open to the public and 34 per cent realising that expert opinions were available from Science Museum staff.

c. Help from museum staff

Table 10.5 gave the percentages of sample members who asked members of staff at each museum for directions. Table 10.7 indicates the proportions of sample members putting a query of any type to a member of staff. It can be seen that far more visitors ask for information at the two South Kensington museums than at the National Railway Museum. Requests from visitors for help finding their way round the museum formed the bulk of queries in the Victoria and Albert Museum and the Science Museum, but were only raised by 29 per cent of the visitors who asked National Railway Museum staff for information (Table

Table 10.7 Percentage of sample asking staff for information

	Victoria and Albert Museum	Science Museum	National Railway Museum
Visitors asking for information	44%	31%	11%
Base	1,003	1,045	596

Source of table: the quota samples of museum leavers.

Table 10.8 The types of information staff were asked for

Visitors asking for:	Victoria and Albert Museum	Science Museum	National Railway Museum
	%	%	%
Directions	83	78	29
Information about a particular exhibit	23	15	42
Other information	12	16	36
Base*	440	323	66

* All visitors asking staff for information.
Percentages add to over 100% as people could give more than one answer.
Source of table: the quota samples of museum leavers.

Table 10.9 How helpfully staff answered visitors' queries

Visitor found staff's answers:	Victoria and Albert Museum	Science Museum	National Railway Museum
	%	%	%
Very helpful	81	81	82
Fairly helpful	14	11	15
Not really helpful	5	9	3
Base	440	323	66

* All visitors asking staff for information.
Source of table: the quota samples of museum leavers.

10.8). The remaining queries at each museum were divided between questions about exhibits and a miscellaneous set of questions including requests for practical information and a few queries about goods in the museum shops.

Sample members appear to have been very satisfied with the response to their queries. As Table 10.9 shows over 80 per cent of questioners found the staff's answers very helpful at each of the three museums.

d. The museum restaurants

The members of the sample who had visited the museum's restaurant—23 per cent of the sample in the Victoria and Albert Museum and about 30 per cent in the other two museums (Table 10.10)—were asked whether they felt that improvements could be made in the way the restaurant was set out and run, or in the quality of the food. As Table 10.11 indicates just over 50 per cent of the sample members using the restaurants of the South Kensington museums saw scope for improvement as did 40 per cent of those members of the National Railway Museum sample who had made use of the restaurant there. (Because of improvements being made to the kitchen facilities the National Railway Museum's restaurant was on minimum service for much of the time quota interviews were being carried out.)

Table 10.12 lists the sort of improvements suggested. Between seven per cent and ten per cent of the visitors at each restaurant saw room for improvement in the quality of food or drink served. A slightly lower percentage at each restaurant said that they would have liked the service to be faster. About ten per cent of the visitors at each South Kensington museum criticised the standard of cleanliness, complaining about

Table 10.10   Percentage of visitors using museum restaurant

	Victoria and Albert Museum	Science Museum	National Railway Museum
Visitors using restaurant	23%	29%	30%
Base	1,003	1,045	596

Source of table: the quota samples of museum leavers.

Table 10.11   Percentage of restaurant users feeling that some improvements could be made

	Victoria and Albert Museum	Science Museum	National Railway Museum
Restaurant users who felt that some improvements could be made	52%	53%	40%
Base	235	308	180

Source of table: the quota samples of museum leavers.

Table 10.12   Percentages of restaurant users suggesting certain specific improvements

Improvements suggested	Victoria and Albert Museum	Science Museum	National Railway Museum
Faster service	6%	6%	8%
Higher standard of cleanliness	9%	10%	2%
More appealing decor	22%	4%	2%
Different location for restaurant	0%	4%	0%
Better quality in food and drink served	10%	7%	10%
Wider range of food	1%	13%	11%
Lower prices	7%	12%	1%
More space to reduce crowding	1%	11%	1%
Better heating or ventilation	6%	3%	0%
Other	17%	16%	18%
Base*	235	308	180

* All restaurant users

The percentages for each museum total to more than the percentages in Table 10.11 since some visitors suggested more than one improvement.

Source of table: the quota samples of museum leavers

uncleared tables and inadequately washed dishes and cutlery. Prices were mentioned by 12 per cent of the people visiting the Science Museum's restaurant and by seven per cent at the Victoria and Albert Museum. At both the Science Museum and the National Railway Museum the most frequent suggestion was that a wider range of food should be served. At the National Railway Museum about half the people making this suggestion asked specifically for more hot food as did about a quarter at the Science Museum. Eleven per cent of visitors to the Science Museum restaurant felt that it was overcrowded and needed more space. Four per cent suggested that it should be located somewhere more accessible than its present position at the back of the museum on the third floor.

These suggestions, and the criticisms which they imply, should not be taken as objective measures of the performance of the restaurants on each of the dimensions concerned but rather as indications of ways in which each could be brought more into line with the tastes and standards of its own particular set of visitors.

This point is illustrated particularly clearly by the comments made by the visitors to the Victoria and Albert Museum's restaurant. By far the most frequent comment from these aesthetically-minded people was the suggestion, made by 22 per cent of the restaurant's visitors, that the restaurant's decor should be improved so that it looked less like a canteen. This should not be taken to imply that the decor of the Victoria and Albert Museum's restaurant was in fact less interesting than the decor of the restaurants at the other two museums. The likely explanation is rather that the Victoria and Albert Museum's visitors are particularly aware of matters of this kind.

e. The museum shops
All three museums contain shops selling a variety of items connected with the museum's collection and the general subject area it covers. At the time of the survey the shops at the Science Museum and the National Railway Museum had recently been rebuilt to new designs. The shop at the Science Museum was visited by 36 per cent of the sample members, while about 60 per cent visited the shops at the other two museums (Table 10.13). Between a third and a half of the people visiting the shops said that they had bought something. At each museum all the members of the sample who had visited the shop were asked whether they felt that improvements could be made in what the shop sold, or in the way it was set out and run. Table 10.14 shows that the proportion of visitors who felt that improvements were possible ranged from 25 per cent at the National Railway Museum to 47 per cent at the Science Museum.

Perhaps the most striking feature of Table 10.15, which lists the improvements suggested, is the contrast between the Victoria and Albert Museum and the Science Museum. The most common suggestion at the Science Museum, made by 36 per cent of the shop's visitors, was that it should stock a wider range of goods. This suggestion was only made by 9 per cent of visitors to the shop at the Victoria and Albert Museum. Their most frequent suggestion, made by 16 per cent, was that the shop should be given more space. Another nine per cent would have liked the lay-out improved. Twenty-five per cent of the people who felt that the range of goods at the Science Museum's shop should be widened suggested that the shop should stock a greater variety of goods generally. However 32 per cent of them asked specifically for a greater variety of books and leaflets; suggestions ranged from guides for the whole museum to explanatory material relating to

Table 10.13   Percentages of all visitors who visited the museum shop

	Victoria and Albert Museum	Science Museum	National Railway Museum
Visitors to museum shop	59%	36%	63%
Base*	1,003	1,045	596

* All visitors.

Source of table: the quota samples of museum leavers.

Table 10.14 Percentages of shop visitors who made purchases or suggesting improvements

	Victoria and Albert Museum	Science Museum	National Railway Museum
Shop visitors making a purchase	36%	42%	48%
Shop visitors suggesting improvements	39%	47%	25%
*Base**	*594*	*375*	*374*

* All shop visitors.

Source of table: the quota samples of museum leavers.

Table 10.15 Percentages of shop visitors suggesting certain specific improvements

Improvements suggested	Victoria and Albert Museum	Science Museum	National Railway Museum
Greater variety of stock	9%	36%	7%
More space	16%	5%	5%
Better lay-out	9%	5%	3%
Making it easier to see the stock	2%	2%	8%
Lower prices	4%	3%	1%
Other	14%	11%	11%
*Base**	*594*	*375*	*374*

* All shop visitors.

The percentages for each museum total to more than the percentages in Table 10.14 since some visitors suggested more than one improvement.

Source of table: the quota samples of museum leavers.

specific galleries, as well as books dealing in general terms with subjects covered by the museum without necessarily focusing on the museum's own collections. Twenty-three per cent wanted more post cards and 21 per cent wanted more models. Other types of goods receiving several mentions were posters (14 per cent) and souvenirs (11 per cent).

The shop at the National Railway Museum attracted comparatively few suggestions relating either to its contents or to any need for greater spaciousness or improved lay-out in general terms. However eight per cent of its visitors, a higher proportion than at the other two museums, thought that something might be done to make it easier to see the stock. There seemed to be a feeling that the way the shop was organised discouraged browsing.

f. Other aspects of the running of each museum

Towards the end of their interviews members of the quota samples of museum leavers were asked whether they felt any improvement was needed in each of a number of specified features of the museum. Their views on one of these features, the provision of signs to help visitors find their way round, have been discussed in Section (a) of this chapter. The percentages feeling that improvements were needed in each of the remaining aspects are listed in Table 10.16.

Table 10.16 Percentage of all visitors who felt improvements were needed in various features of each museum

Visitors feeling improvements were needed in:	Victoria and Albert Museum	Science Museum	National Railway Museum
Seating	27%	26%	14%
Heating and air-conditioning	15%	27%	8%
Lighting	33%	9%	8%
Opening and closing times	22%	17%	14%
Other features	19%	19%	21%
Base	*1,003*	*1,045*	*596*

Source of table: the quota samples of museum leavers.

i. Seating

Just over a quarter of the sample at each South Kensington museum felt that the museum's seating needed to be improved. At the National Railway Museum the demand for better seating provision was less than at the other two museums. Nevertheless seating provision was, equally with the museum's hours of business, one of the two features most commonly felt to need improvement at the National Railway Museum. At all three museums the improvement wanted was simply the provision of more seats. Not surprisingly, Table 10.17 shows that the proportion of visitors seeing a need for more seating was highest among those who had been in the museums longest. In the two South Kensington museums older visitors were, as might be expected, rather more likely than younger ones to think more seats should be provided. It is striking, however, that substantial proportions among the youngest members of the sample and among those who had spent least time in the museum also felt that the seating should be improved.

Table 10.17 Percentage who felt the provision of seating needed to be improved, by age and time spent in museum

	Victoria and Albert Museum (a)	Science Museum (b)	National Railway Museum (c)	Bases (a)	(b)	(c)
Age						
11–20	25%	21%	16%	*255*	*337*	*187*
21–30	22%	28%	10%	*262*	*320*	*198*
31–40	32%	27%	17%	*131*	*167*	*82*
41–50	30%	33%	17%	*114*	*110*	*48*
51 and over	28%	32%	14%	*241*	*111*	*81*
Time in museum						
Less than 1 hour	23%	22%	12%	*224*	*171*	*178*
1 but less than 2 hours	25%	24%	14%	*472*	*435*	*309*
2 but less than 3 hours	31%	28%	15%	*211*	*313*	*93*
3 hours or more	34%	37%	20%	*94*	*126*	*15*

Source of table: the quota samples of museum leavers.

ii. Heating and air-conditioning

One might expect the functioning of the heating and air-conditioning system at each museum (and therefore visitors' views on the subject) to depend on the weather and thus on the time of year. This means that some caution is needed in interpreting Table 10.18 and the second row of Table 10.16. First because, although the days on which there were quota sample interviews with museum leavers were spread through the whole survey year, no explicit attempt was made to ensure that the proportion of the quota sample that was interviewed at each time of year matched the proportion of the museum's visitors who came at that time of year. A second reason is that, for organisational reasons, the days selected for quota sample interviews tended to be bunched together. Thus in the case of the Victoria and Albert Museum and the Science Museum the interviews contributing to the 'December to February' column in Table 10.18 were all conducted during one week in January, while those at the National Railway Museum all took place in the space of four days in February. Though the interviewing during other seasons was not so concentrated, it is necessary to bear in mind that the figures for particular seasons may have been influenced by spells of untypical weather on a few days.

Even with these qualifications two major conclusions emerge from Table 10.18. The first is that at all three museums, for by far the greater part of the year, more visitors feel too hot than too cold. The second is that the proportions of Science Museum visitors who felt too hot was not constant throughout the year. Although the variation in this proportion—from nine per cent of visitors interviewed in the months December to February to 29 per cent of visitors interviewed between March and May—reflects conditions on the particular days on which interviewing took place and

cannot be taken as an accurate representation of conditions over each three month period, the figures do indicate that variations in the weather were not entirely cancelled out by the operation of the museum's heating and ventilation system.

iii. Lighting

Thirty-three per cent of the sample at the Victoria and Albert Museum thought that the museum's lighting needed improvement, compared with less than ten per cent at the other two museums. The main source of dissatisfaction at the Victoria and Albert Museum was the lighting of the exhibits or of specific galleries, mentioned by 22 per cent of the sample. Among the other points mentioned were reflections on glass (four per cent of the sample), poor lighting on labels (three per cent) and the level of lighting overall (three per cent). Lighting has already been discussed in the chapters on presentation, the figures quoted here merely confirm that a considerable proportion of the museum's visitors see it as a problem.

iv. Opening and closing times

The proportions of the sample shown by Table 10.16 to feel that a change was needed in the museum's hours of business ranged from 14 per cent in the case of the National Railway Museum to 22 per cent in the case of the Victoria and Albert Museum. Table 10.19 sets out the ways in which it was felt each museum's hours should change. Alone of the three museums, the Victoria and Albert Museum closed on Fridays during the survey year. Seven per cent of the sample felt that it should remain open. During the survey year the Sunday opening hours of all three museums were from 2.30 pm to 6 pm. Nine per cent of the Victoria and Albert Museum's visitors wanted the museum to open earlier on Sundays, as did similar proportions of the sample at the other two museums. This suggestion

Table 10.18 Percentages of each museum's visitors feeling that the heating and air-conditioning system needed improvement for various reasons, by time of year

Visitors finding each museum's heating/air-conditioning:	Time of year of visit				
	December to February	March to May	June to August	September to November	All visitors
Victoria and Albert Museum					
Too hot	8%	13%	9%	7%	10%
Too stuffy	4%	4%	4%	1%	3%
Too cold	3%	1%	1%	5%	3%
Other	2%	1%	1%	0%	1%
Any of the above	13%	16%	15%	14%	15%
Base	*98*	*337*	*273*	*295*	*1,003*
Science Museum					
Too hot	9%	29%	22%	16%	20%
Too stuffy	2%	9%	4%	4%	5%
Too cold	2%	0%	1%	2%	2%
Other	0%	2%	2%	0%	1%
Any of the above	12%	38%	28%	22%	27%
Base	*123*	*255*	*315*	*352*	*1,045*
National Railway Museum					
Too hot	13%	4%	9%	3%	6%
Too stuffy	3%	1%	5%	0%	2%
Too cold	0%	2%	0%	1%	1%
Other	0%	0%	0%	0%	0%
Any of the above	13%	8%	11%	3%	8%
Base	*72*	*187*	*184*	*153*	*596*

Source of table: the quota samples of museum leavers.

was made particularly by Sunday visitors. Table 10.20 shows that over a third of all the members of the sample visiting the two South Kensington museums on a Sunday thought that the Sunday opening hours were too short, an opinion shared by over a quarter at the National Railway Museum.

Table 10.19   Percentage of the sample who felt that various changes were needed in the museums' opening and closing times

Visitors feeling that the museum should:	Victoria and Albert Museum	Science Museum	National Railway Museum
Open on Friday	7%	—	—
Open earlier on Sunday	9%	8%	7%
Open earlier generally	2%	4%	5%
Remain open later	5%	6%	2%
Other	1%	1%	1%
Base	*1,003*	*1,045*	*596*

Visitors could suggest more than one change.

Source of table: the quota samples of museum leavers.

Table 10.20   Percentage of Sunday visitors who thought the museum needed to open earlier on Sundays

	Victoria and Albert Museum	Science Museum	National Railway Museum
Sunday visitors who thought the museum needed to open earlier on Sundays	39%	37%	29%
Base	*168*	*168*	*96*

Source of table: the quota samples of museum leavers.

Table 10.21   Percentage thinking the museum needed to open earlier generally, by time of interview

Time interview started	Victoria and Albert Museum	Science Museum	National Railway Museum
Before 12 o'clock	7%	10%	12%
Base	*227*	*183*	*206*
12 to 12.59	1%	7%	2%
Base	*141*	*139*	*84*
1 pm or later	1%	2%	1%
Base	*633*	*723*	*306*

Source of table: the quota samples of museum leavers.

A substantial proportion of the quota sample interviews (23 per cent at the Victoria and Albert Museum, 18 per cent at the Science Museum, 35 per cent at the National Railway Museum) were conducted with people who finished their visit before 12 o'clock. Comparable figures for interviews obtained from the count-based samples (excluding under-elevens and visitors in organised parties) are five per cent, five per cent and ten per cent, so that it can be seen that the quota samples over-represent early leavers by a factor of more than three. This affects the interpretation of the next line in Table 10.19 which shows the proportion of the quota sample at each museum who felt the museum should open earlier, a proportion which ranges from two per cent at the Victoria and Albert Museum to five per cent at the National Railway Museum. Table 10.21 suggests that the proportion of all visitors who hold this view at each museum is probably much lower,

since the view was much more frequently expressed by people leaving each museum before 12 o'clock than by people leaving in the afternoon. The figures in Table 10.19 make it appear more widespread than it is because they over-represent early leavers.

The fourth row of Table 10.19 shows the proportions of quota sample visitors who thought each museum should stay open later. As this opinion was not confined to people who left late in the day the figures in the fourth row are not likely to have been greatly distorted by the timing of the quota interviews.

v. Other features felt to need improvement

After giving their views on the features of the museum specified by the interviewers visitors were asked whether they thought improvements were needed in any other aspect of the museum. Table 10.16 shows that about 20 per cent of the visitors interviewed at each museum thought that other improvements were needed. Table 10.22 lists the improvements most frequently suggested and gives the proportion of the whole sample mentioning each.

Table 10.22   Percentage of the sample who felt that improvements were needed in various other aspects of the museum

Improvements felt to be necessary	Victoria and Albert Museum	Science Museum	National Railway Museum
Better labelling	3%	2%	2%
Better presentation	3%	2%	5%
Visitors should be allowed to climb onto exhibits	0%	0%	5%
Difference in the choice of exhibits	1%	2%	4%
Galleries should not be shut	2%	0%	0%
Better upkeep of working models telephones, lifts and escalators	0%	4%	0%
Base	*1,003*	*1,045*	*596*

Many of the comments made in answer to this question could not be fitted into any simple category. For this reason the table excludes additional comments made by 102 individuals at the Victoria and Albert Museum, 114 at the Science Museum and 58 at the National Railway Museum.

Source of table: the quota samples of museum leavers.

At the Victoria and Albert Museum labelling and presentation attracted most comments (three per cent of the sample each), followed by the suggestion that individual galleries should not be closed. At the Science Museum four per cent of the sample felt that the upkeep of working models, telephones, lifts or escalators needed to be improved. The most frequent comments at the National Railway Museum were suggestions that aspects of presentation should be improved and the specific suggestion that some exhibits should be provided for visitors to climb onto (five per cent of the sample making each suggestion). A number of railway buffs felt that there should have been more examples of the objects of their own particular enthusiasms on show (four per cent).

These comments were made in response to an open question at the end of a long series of questions relating to the way the museum was run. The proportion of visitors mentioning at this point that a specific type of improvement was needed is therefore likely to be considerably lower than it would have been if the improvement concerned had been mentioned specifically in the question. In effect all Table 10.22 does is to provide hints of more areas in which changes might be appreciated.

11 Special categories of visitors

a. General

Three kinds of visitor presented particular problems from the viewpoint of the survey. These were children aged ten or less, children visiting in school parties and foreign visitors. The specific problems differed. With under-elevens it was the difficulty of conducting detailed interviews with such young children. In the case of school parties it was the impossibility of detaining individual children leaving the museums with school parties for long enough to carry out the comparatively lengthy quota interviews. Language problems meant that information could not be obtained from all overseas visitors. The following sections deal in turn with these three groups.

b. Under-elevens

Children aged ten or less were not included in the quota samples of people leaving the museums or selected galleries inside them. Instead we attempted to obtain some information about their reactions by putting questions to their parents. Though some of the children may have expressed their own views at this point, Tables 11.1 and 11.2 record parents' views of their children's reactions rather than the views of the children themselves. The data is only valid to the extent that the parents were accurate observers of their children's reactions. The tables are based on interviews at the Science and National Railway Museums only since, as we saw in Chapter 2, the Victoria and Albert Museum received comparatively few visits from young children, too few to provide adequate numbers in our sample to analyse at all reliably.

Table 11.1 Parents' judgements of which galleries at the Science Museum their young children found most interesting

Gallery and floor*	Parents visiting with children aged 0–10 only
	%
Children's gallery (LG)	28
Rail Transport and Trams (G)	8
Exploration (G)	25
Temporary Exhibition gallery on ground floor	5
Temporary Exhibition gallery on first floor (Challenge of the Chip)	5
Computing Then and Now (2)	3
Aeronautics (3)	8
Other or no answer	19
Base	*109*

*See note on Table 8.12

Source of table: the quota samples of museum leavers.

Parents leaving each museum were asked which section of the museum they thought their children had found most interesting. Table 11.1 shows how they replied at the Science Museum. Since most visitors to the Science Museum only stop to look at a few of the galleries the figures in the table inevitably reflect the decisions of the children, or their parents, about which galleries to visit as well as the qualities of the galleries themselves.

The remarkable thing about the list of permanent galleries which parents said that their children found most interesting is that it consists of the same five 'star' galleries (or subjects) which played the largest part in attracting all visitors to the museum, which all visitors were particularly likely to visit and which were given the highest interest ratings by the sample generally. It seems that these subjects or galleries—the Children's gallery itself, space travel (represented by the Exploration gallery), road and rail transport, aeronautics and computers—exert their attraction virtually regardless of age.

This statement should perhaps be qualified a little in the case of the Children's gallery. The appeal for its adult visitors may well lie mainly in the enjoyment of the children themselves. The large number of children who found the gallery interesting—28 per cent of the parents contributing to Table 11.1 thought that it was the gallery which their children had found most interesting—were enjoying it on their own account.

Parents leaving the selected galleries at the two technical museums were asked the following question about the gallery's design:

"Do you feel that [the particular gallery] *is well presented from the point of view of your children, or do you feel that it wasn't really designed with children like yours in mind?"*

Table 11.2 gives their answers. It is instructive to compare Table 11.2 with Table 6.1 in Chapter 6 which sets out the answers visitors gave when asked whether they themselves had found each gallery attractive. The three selected galleries which visitors found most attractive on their own account at the two technical museums—the Aeronautics and Exploration galleries at the Science Museum and the Main Hall at the National Railway Museum—were also the three galleries which parents thought best presented from the point of view of their children. Here again, if the parents are to be believed, the tastes of young children seem to mirror those of older visitors.

Table 11.2 Parents' judgements of how well the selected galleries were presented from the point of view of their young children

Parents' judgement	Science Museum				National Railway Museum	
	Aeronautics	Printing and Paper	Time Measurement	Exploration	Balcony	Main Hall
	%	%	%	%	%	%
Well presented from their children's point of view	75	40	31	72	42	69
Not really designed with children like theirs in mind	25	60	69	28	58	31
Base	*20*	*20*	*13*	*30*	*34*	*35*

The table is based on the comments of parents visiting with their own children all of whom were aged 10 or less.

Source of table: the quota samples interviewed as they left selected galleries.

Table 11.3 Proportion of 11–20 year old school party visitors who had been told by their teacher to visit selected galleries

	Victoria and Albert Museum				Science Museum				National Railway Museum
	Art of China and Japan	British Sculpture	Tudor Art	Continental 17th Century Art	Aeronautics	Printing and Paper	Time Measurement	Exploration	Balcony
Percentage told to visit the gallery by their teacher	18%	21%	25%	15%	17%	6%	12%	21%	77%
Base	*12*	*14*	*16*	*13*	*32*	*32*	*25*	*28*	*14*

Source of table: the quota samples interviewed as they left selected galleries.

c. School parties

It was possible to include people visiting with school parties in the count-based samples interviewed as they left each museum and results derived from this source have been presented in Chapters 2, 3 and 8. They were not included in the quota samples of museum leavers as individuals leaving with a group could not be expected to spare the necessary time. However we did hope to gather some information about their viewing patterns inside each museum by including them in the quota samples for each of the ten selected galleries. The hope was that these school children—aged 11 or more—would have been instructed by their teachers to visit the galleries where they were interviewed, and that it would be possible to say something about the way such 'directed' visits worked in practice.

However it turned out, as Table 11.3 shows, that very few of the school children we spoke to had been directed to visit the gallery in question—either because they had not been told to visit any particular part of the museum or because the galleries selected for interviewing were not ones which they had been directed to look at. It may well be that school children who had been directed to study a particular gallery were harder for interviewers to intercept than school children who had simply wandered in and out, and that as a result the quota samples may include a disproportionate number of the latter. Whatever the reason, the data does not provide the type of information we had hoped for.

d. Overseas visitors

Chapter 2 (Table 2.10) gave figures—derived from the weighted results from the count-based samples of museum leavers—for the proportion of each museum's visitors who were resident overseas. These figures—30 per cent in the case of the Victoria and Albert Museum and 19 and nine per cent respectively at the Science and National Railway Museums—were based on the replies of the visitors whom we succeeded in interviewing. However, as is shown in Appendix 1 (Table A4) a number of visitors could not be interviewed at all because neither they nor any companion spoke enough English. Since most of these people must have been foreign residents our figures are likely to understate the total number of foreign visitors and the proportion of those visitors who did not speak English. It is possible that the problem affects residents of some countries more than others because of differences in the standard to which they learn English. The following results relate only to those visitors we were able to interview.

Table 11.4 gives the proportions of overseas visitors to each museum whose first language was English. This was assumed without asking in the case of residents of the Irish Republic, Australia, New Zealand and the USA. Citizens of all other foreign countries were asked to state their first language. Table 11.5 looks at the same topic in another way. It takes the foreign visitors whose first language was not English and shows what proportion they formed of the sample interviewed at each museum. They turn out to be much more numerous at the two London museums than in the National Railway Museum. Table 11.6 lists the first languages of overseas visitors who were not native English speakers at the two London museums. Table 11.7 shows that overseas visitors, particularly those who were not native English speakers, were generally making their first visit to the museum in question.

Table 11.4   Proportions of native English speakers among the overseas visitors interviewed at each museum

	Victoria and Albert Museum	Science Museum	National Railway Museum
Overseas visitors whose first language is English*	54%	39%	62%
Weighted base	300	190	90

* *It was assumed that residents of the Irish Republic, Australia, New Zealand and the USA were native English speakers. Visitors from all other foreign countries were asked to state their first language.*

Source of table: weighted results from the count-based samples of museum leavers.

Table 11.5   Foreign visitors whose first language was not English as a proportion of all visitors interviewed at each museum

	Victoria and Albert Museum	Science Museum	National Railway Museum
Foreign visitors whose first language was not English	14%	11%	3%
Weighted base*	1,000	1,000	1,000

* *All visitors*

Source of table: weighted results from the count-based samples of museum leavers.

Our only measure of how much foreign visitors enjoyed their visit overall comes from a question put to the quota samples of museum leavers (the overall results for each museum's visiting public will be discussed in Chapter 12). Table 11.8 compares the answers given by British residents with those given by overseas visitors who spoke English as their first language and by overseas visitors who were not native English speakers. People in this last group were only included in the quota samples if their English was good enough for the comparatively lengthy quota interviews. The results for this group, and for overseas visitors who spoke English as their first language, are encouraging. Both categories of overseas visitors seem to have found the museums at least as enjoyable as did their British counterparts.

Table 11.6   First languages of foreign visitors who were not native English speakers

First language	Victoria and Albert Museum	Science Museum
	%	%
Swedish/Danish/Norwegian	12	13
German	21	21
Dutch	7	16
French	14	17
Italian	11	4
Spanish	6	5
Portuguese	2	7
Other European languages	9	2
Hebrew	4	0
Arabic	2	6
Chinese	3	3
Other non-European languages	9	6
Weighted base	140	110

Source of table: weighted results from the count-based samples of museum leavers.

Table 11.7   Proportions of British and overseas visitors who had been to each museum before

	Victoria and Albert Museum	Science Museum	National Railway Museum
British residents	65%	60%	38%
Base	700	810	910
Overseas residents speaking English as their first language	35%	33%	11%
Base	160	70	50
Overseas residents whose first language was not English	14%	23%	7%
Base	140	110	30

Source of table: weighted results from the count-based samples of museum leavers.

Table 11.8   Proportions of British and overseas visitors rating each museum very enjoyable

	Victoria and Albert Museum	Science Museum	National Railway Museum
British residents	52%	57%	71%
Base	651	797	526
Overseas residents speaking English as their first language	67%	65%	77%
Base	252	112	43
Overseas residents whose first language was not English	63%	58%	71%
Base	98	120	24

Source of table: the quota samples of museum leavers.

12　The visit as a whole

a. Enjoying the visit

Visitors selected for the quota samples as they left each museum were asked whether their visit had been very enjoyable, fairly enjoyable or not really enjoyable. Table 12.1 sets out their answers. A majority of each museum's visitors had found their visit very enjoyable, visitors to the National Railway Museum—71 per cent of whom rated it very enjoyable—being the most appreciative.

It was noted in Chapter 5 that people's enjoyment of particular galleries in each museum was closely related to how interesting they found them. One would therefore expect people's enjoyment of the museum as a whole to be closely related to their interest in the galleries they had looked at during their visit. Figure 12.1 and Table 12.2 confirm that this is so. The table is based on the interest level ratings that visitors gave each of the galleries where they stopped to look at something during their visit. The proportion of visitors who found their visit to the Victoria and Albert Museum very enjoyable rises from 12 per cent in the case of visitors who gave a high rating to less than ten per cent of the galleries they had looked at, to over 80 per cent of those who had thought 70 per cent or more of the galleries very interesting.

At the Science Museum the proportion of visitors who felt their visit had been very enjoyable rises from 13 per cent of those visitors who rated fewest galleries highly to 75 per cent in the case of people who had found nearly all the galleries they looked at very interesting.

Table 12.1　Enjoyment ratings at each museum

The visit was:	Victoria and Albert Museum	Science Museum	National Railway Museum
	%	%	%
Very enjoyable	57	59	71
Fairly enjoyable	40	40	28
Not really enjoyable	3	2	1
Base	*1,003*	*1,045*	*596*

Source of table: the quota samples of museum leavers.

Figure 12.1 Interest in Science Museum galleries and enjoyment of the museum as a whole (from Table 12.2)

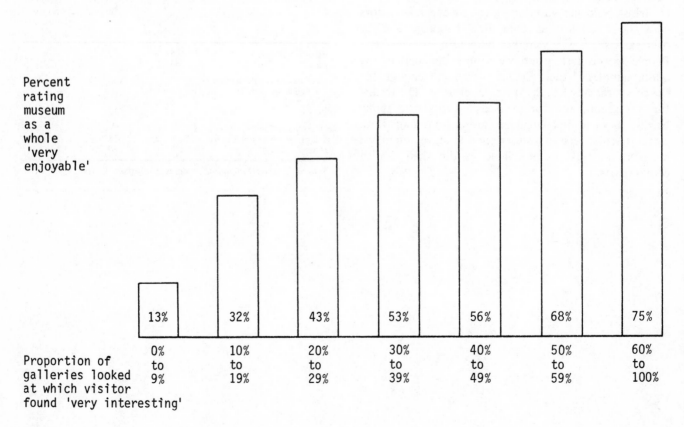

Percent rating museum as a whole 'very enjoyable'

| 13% | 32% | 43% | 53% | 56% | 68% | 75% |

| 0% to 9% | 10% to 19% | 20% to 29% | 30% to 39% | 40% to 49% | 50% to 59% | 60% to 100% |

Proportion of galleries looked at which visitor found 'very interesting'

The kinds of people who were particularly likely to find individual galleries interesting also tended to enjoy their whole visit more than others. Table 12.3 shows that enjoyment is strongly related to age with older people more likely to report a 'very enjoyable' visit. Table 12.4 looks at how people's enjoyment of their visit related to the importance they attached to the educational side of museum visiting. Visitors for whom it was important to feel that they had learnt something at the museum reported higher enjoyment levels than visitors who 'just liked looking at things' though still a majority of the latter said they enjoyed themselves very much. This parallels the greater interest that 'learners' reported in individual galleries.

Table 12.2 Proportions of visitors rating each London museum 'very enjoyable' by the proportion of galleries they rated 'very interesting'

Proportions of galleries rated 'very interesting'	Victoria and Albert Museum (a)	Science Museum (b)	Bases (a)	(b)
	% rating museum 'very enjoyable'			
Less than 10 per cent	12%	13%	84	68
10 but less than 20 per cent	29%	32%	24	50
20 but less than 30 per cent	26%	43%	77	98
30 but less than 40 per cent	49%	53%	94	144
40 but less than 50 per cent	48%	56%	96	152
50 but less than 60 per cent	58%	68%	213	200
60 but less than 70 per cent	66%	75%	112	107
70 but less than 80 per cent	82%	75%	57	61
80 but less than 90 per cent	88%	74%	48	43
90 to 100 per cent	80%	75%	192	119

Source of table: the quota samples of museum leavers.

Table 12.3 Proportions of visitors rating each museum 'very enjoyable' by age

Age	Victoria and Albert Museum (a)	Science Museum (b)	National Railway Museum (c)	Bases (a)	(b)	(c)
	% rating museum 'very enjoyable'					
11–20	44%	45%	58%	255	337	187
21–30	46%	55%	66%	262	320	198
31–40	57%	67%	82%	131	167	82
41 and over	75%	80%	91%	355	221	129

Source of table: the quota samples of museum leavers.

Table 12.4 Proportions of visitors rating each museum 'very enjoyable' by attitude to learning

Attitude to learning	Victoria and Albert Museum	Science Museum	National Railway Museum
	% rating museum 'very enjoyable'		
Important to feel you have learnt something	61%	63%	77%
Base	*542*	*583*	*261*
Just like looking at things	52%	52%	67%
Base	*445*	*431*	*330*

Source of table: the quota samples of museum leavers.

Table 12.5 Proportions of visitors rating each museum 'very enjoyable' by sex

	Victoria and Albert Museum (a)	Science Museum (b)	National Railway Museum (c)	Bases (a)	(b)	(c)
	% rating museum 'very enjoyable'					
Males	52%	62%	72%	458	587	324
Females	61%	54%	70%	545	458	272

Source of table: the quota samples of museum leavers.

Table 12.5 indicates that women found the Victoria and Albert Museum slightly more enjoyable than did men, while men and boys gave slightly higher ratings to the two technical museums.

This is consistent with the general drift of results in Chapters 2, 3 and 7. However the main message of Table 12.5 is not that men and women differ in their assessment of the museums but that their judgements are in fact rather similar, despite the existence of some differences in the initial enthusiasm and degree of technical knowledge that men and women bring to the two technical museums.

In Chapter 10 it was suggested that various practical aspects of the way the museums were run might make an important contribution to visitors' enjoyment of their visits. The chapter reported responses to a series of questions, put to all members of the quota sample of people leaving each museum, about whether they felt improvements needed to be made in the museum's seating provision, heating and air-conditioning system, lighting, signposting, hours of opening or any other feature of the museum. In all visitors were given six opportunities to say that, yes, improvements were needed. The number of times a visitor said that improvements were needed provides an index of how dissatisfied they were with these aspects of the practical organisation of the museum. (Note that this index does not cover all aspects of the museum's organisation. It specifically excludes the views on the organisation of the shop and restaurant of those visitors who used them.)

Table 12.6 shows how the number of improvements that visitors thought necessary was related to their enjoyment of each museum. If what visitors saw as deficiencies in these practical aspects of the museums' organisation had greatly affected their enjoyment one would expect those who saw considerable need for improvement to report lower levels of enjoyment than those who did not. In fact the figures show little relationship between enjoyment of the visit and satisfaction with these aspects of the museum's organisation. They certainly seem to have been much less important for visitors' enjoyment than were the interest sustaining qualities of the galleries themselves.

The quality of the signposting has a bearing on the interest visitors derive from the galleries they visit— since it affects their ability to locate the galleries (or, in the National Railway Museum, the parts of the Main Hall) of most interest to them. Table 12.7 suggests that this particular aspect of the museums' practical organisation does have some effect on visitors' enjoyment. Table 12.7 indicates that in all three museums visitors who were dissatisfied with the signs provided to help them find their way round reported slightly lower levels of enjoyment than people who felt the signposting needed no improvement.

Museums do not exist simply to provide enjoyment. Their exhibitions are also intended to have an educational function—widening their visitors' intellectual horizons and deepening their appreciation and understanding of the exhibits displayed. Fears are sometimes expressed that efforts to make museums enjoyable may be at the expense of their educational role.

A good deal of what has gone before in this report suggests that the two objectives do not conflict as often as one might suppose. Enjoyment is closely connected to interest and many of the aspects of presentation which enhance interest do so because they make it easier for the visitor to appreciate the exhibits that are displayed. In fact, visitors' enjoyment and interest themselves assist in the museums' educational functions. People who are interested in what they see are willing to spend more time looking at it. In Chapter 5 we saw that visitors who found each gallery very interesting tended to spend more time looking round it than those who did not. Tables 12.8 and 12.9 repeat this finding at the level of the whole museum. Visitors who found each museum very enjoyable tended to spend a slightly longer time on their visit than other visitors. New visitors to each museum were considerably more likely to want to return if they had found their first visit enjoyable. It seems that enjoyment of what they see influences visitors to give the museum more attention both on their current visit and by returning to look at its contents again.

Table 12.6 Proportions of visitors rating each museum 'very enjoyable' by the number of improvements they thought were needed in its practical organisation

Number of improvements thought to be necessary	Victoria and Albert Museum (a)	Science Museum (b)	National Railway Museum (c)	Bases (a)	(b)	(c)
	% rating museum 'very enjoyable'					
None	59%	61%	76%	163	243	285
1	57%	62%	66%	336	369	200
2	52%	55%	71%	273	282	84
3 or more	61%	54%	56%	231	151	27

Source of table: the quota samples of museum leavers.

Table 12.7 Proportions of visitors rating each museum 'very enjoyable' by whether they thought improvements were needed in the signs provided to help people find their way round

Visitor thought that:	Victoria and Albert Museum (a)	Science Museum (b)	National Railway Museum (c)	Bases (a)	(b)	(c)
	% rating museum 'very enjoyable'					
Improvements to the signs were needed	52%	55%	64%	516	416	73
Improvements to the signs were not needed	62%	61%	73%	450	605	490
No opinion	59%	59%	61%	34	22	33

Source of table: the quota samples of museum leavers.

b. Time and tiredness

Table 12.10, derived from the count-based samples, shows the length of time visitors spent at each museum. Visitors to the Science Museum tended to spend most time on their visits—37 per cent taking two hours or more. Visitors to the National Railway Museum spent

Table 12.8 Time spent in each museum by enjoyment rating

Time spent	Victoria and Albert Museum		Science Museum		National Railway Museum	
	Very enjoyable	Fairly or not really enjoyable	Very enjoyable	Fairly or not really enjoyable	Very enjoyable	Fairly or not really enjoyable
	%	%	%	%	%	%
Less than 1 hour	15	33	13	22	25	42
1 hour but less than 2 hrs	45	50	39	45	53	49
2 hours but less than 3 hrs	26	14	33	25	19	8
3 hours or more	14	3	15	8	3	1
Base	*571*	*431*	*612*	*431*	*424*	*172*

Source of table: the quota samples of museum leavers.

Table 12.9 Whether first-time visitors wanted to return to see more of the museum by enjoyment of visit

	Victoria and Albert Museum		Science Museum		National Railway Museum	
	Very enjoyable	Fairly or not really enjoyable	Very enjoyable	Fairly or not really enjoyable	Very enjoyable	Fairly or not really enjoyable
	%	%	%	%	%	%
Wanted to come back and see more	92	70	89	58	71	35
Felt they had seen as much as they wanted to	8	30	11	42	29	65
Base	*254*	*214*	*229*	*148*	*288*	*98*

Source of table: the quota samples of museum leavers.

least time, only 17 per cent extended their visit for more than two hours. However the most striking feature of the table is not that there is some difference between the amounts of time spent at each museum but rather that the differences are so slight. Although the two London museums contain far more galleries and exhibits than the National Railway Museum, the time their visitors spend looking round them is only slightly longer than the time its visitors devote to the National Railway Museum. In all three museums the most common length for a visit is between one and two hours—visits of this length account for just over 40 per cent of cases at the two London museums and slightly more than half the visits to the National Railway Museum.

Table 12.10   Time spent in each museum

Time spent	Victoria and Albert Museum	Science Museum	National Railway Museum
	%	%	%
Less than 1 hour	31	22	29
1 hour but less than 2 hours	43	41	53
2 hours but less than 3 hours	18	25	14
3 hours or more	8	12	3
Weighted base	1,000	1,000	1,000

Source of table: weighted results from the count-based samples of museum leavers.

It seems that, given museums as large as even the smallest of these three, the extent of the museum itself is not the major factor determining the amount of time that people spend on their visits. An attempt to explain the pattern shown in Table 12.10 must point to other determining factors.

One such factor may be the visitors' stamina, both physical and mental. The following question was put to members of the quota samples of people leaving each museum:

"Some people find that going round a museum can be tiring. Now that you are leaving do you feel very tired, fairly tired or not really tired?"

Tables 12.11 and 12.12 show how the pattern of replies varied according to the length of time that visitors had spent in each of the two London museums.

The results confirm that museum visiting is a tiring activity. At the two London museums the proportion of visitors who on leaving felt very tired rises from less than ten per cent in the case of people who had spent less than an hour in the museum to about a quarter of the visitors who had spent three hours or more in the museum. The proportion of visitors who reported themselves to be 'not really tired' falls from over two thirds of the people who left within an hour of their arrival to less than one third of those who had spent over three hours in either museum.

The pattern at the National Railway Museum, given in Table 12.13, is rather different. Although the proportion who reported some degree of fatigue does increase with the length of the visit, the percentage who report

themselves very or fairly tired is much lower than in the case of people who spent a similar length of time in either the Victoria and Albert Museum or the Science Museum. For instance only 31 per cent of the visitors who had spent between two and three hours at the National Railway Museum admitted to feeling tired, compared to 61 per cent of those who spent that long in the Victoria and Albert Museum and 63 per cent in the Science Museum.

Among the reasons for this difference between the National Railway Museum and the other two museums may be the fact that the National Railway Museum contains fewer staircases and corridors than the other two. It may also be that visits to the National Railway Museum take place in a relaxed, untiring atmosphere—we noted in Chapter 2 that it was very much a museum for family outings.

Though tiredness may limit the length of the visit, it does not appear to affect visitors' enjoyment of the museums. Table 12.14 shows no tendency for visitors who felt tired to report less enjoyment than those who still felt comparatively fresh. It seems likely that visitors to a museum who begin to feel too tired to enjoy spending more time there decide to leave before tiredness can spoil much of the visit.

Table 12.11   Tiredness by time spent in the Victoria and Albert Museum

	Less than 1 hour	1 hour but less than 2 hours	2 hours but less than 3 hours	3 hours or more
	%	%	%	%
Very tired	7	9	17	23
Fairly tired	15	34	44	45
Not really tired	78	57	39	32
Base	224	472	211	94

Source of table: the quota samples of museum leavers.

Table 12.12   Tiredness by time spent in the Science Museum

	Less than 1 hour	1 hour but less than 2 hours	2 hours but less than 3 hours	3 hours or more
	%	%	%	%
Very tired	9	16	18	28
Fairly tired	22	35	45	48
Not really tired	68	49	37	24
Base	171	435	313	126

Source of table: the quota samples of museum leavers.

Table 12.13   Tiredness by time spent in the National Railway Museum

	Less than 1 hour	1 hour but less than 2 hours	2 hours but less than 3 hours	3 hours or more
	%	%	%	%
Very tired	2	2	2	0
Fairly tired	8	22	29	33
Not really tired	89	76	69	67
Base	178	309	93	15

Source of table: the quota samples of museum leavers.

Table 12.14 Proportions of visitors rating each museum 'very enjoyable' by tiredness

	Victoria and Albert Museum (a)	Science Museum (b)	National Railway Museum (c)	Bases (a)	(b)	(c)
	% saying 'very enjoyable'					
Very tired	67%	58%	–	116	177	12
Fairly tired	57%	62%	70%	330	392	114
Not really tired	55%	57%	72%	557	474	470

Source of table: the quota samples of museum leavers.

Appendix 1   Selecting samples of visitors

The main outlines of the sampling scheme were set out in Chapter 1. Briefly, three different kinds of sample were taken at each museum:

A count-based sample of people leaving the museum.

A quota sample of people leaving the museum.

Quota samples of people leaving a few selected galleries within the museum.

After being weighted to allow for differing overall chances of being interviewed on different days and to correct for non-response, the figures from the count-based sample show what proportion of visits over the survey year were made by different categories of visitor—young or old, first-time or return visitors etc—and the reasons for which these visits were made. The figures derived from the quota samples cannot be used to describe the visiting public. The purpose of the quota-based interviews was to obtain visitors' reactions to the museums and to relate these to the visitors' own characteristics and interests.

This appendix describes the 'mechanics' involved in selecting the samples and weighting the figures. It concludes with a short discussion of the reliability of the survey's results.

The count-based samples
The sample for each museum was drawn in such a way that each visitor (strictly each visit) made the same potential contribution to the results. The chance of being interviewed varied according to the date of the visit, but this was corrected for in the weighted figures by giving greater weight to the replies of people who visited on days when there was a lower overall chance of selection.

Selecting the days
Objectives
At each museum interviewing took place on 16 days spread through the survey year (3 December 1979—2 December 1980). It was important that the samples selected should be representative both of the different days of the week and of all times of the year. The key points regarding days of the week were:

To keep a balance between weekdays and weekend days (Saturdays, Sundays, bank holidays)

To keep a balance between Saturdays and Sundays

To ensure that not more than one bank holiday was selected (since bank holidays were likely to be highly untypical)

To spread the weekday interviewing among different days of the week.

The general objectives regarding times of year were:

To spread interviewing throughout the year

To keep a balance between term-time and holidays

To limit the amount of interviewing done in the middle of the winter when it was thought that attendance patterns might be untypical.

Two further points affecting the selection of term-time weekdays were:

Both halves of each school term should be represented

Taking the three terms together, there should be a balance between interviews near the start of term, interviews in the middle of term, and interviews towards the end of term.

The constraints listed above went a long way toward determining the probability of choosing particular days for the sample. However, within the limits imposed, it was decided to give a higher chance of selection to days when the museums were expected to be busy. Doing so would reduce the extent to which data from these samples would need to be weighted.

Main features of the procedure
Weekdays and weekend days were selected separately. However the basic features of the two schemes were the same.

1 The year was divided into segments—one day from each segment being included in the sample.

2 The probability of choosing each specific date, conditional on the choice in the segment concerned falling on the relevant day of the week, was then set.

3 There was a procedure to decide which day of the week was to be selected in each segment—that is, in the case of weekdays whether it was to be a Monday, Tuesday, Wednesday, Thursday or Friday. In the case of weekend days the choice was between Saturdays and Sundays or, in the case of segments containing a bank holiday, first between Saturdays and Sundays/bank holidays and then between a procedure that could only produce a Sunday and a procedure that might produce a Sunday or might produce a bank holiday.

4 There was a procedure for selecting the actual date on which interviewing would take place, once the day of the week selected for each segment was known. This procedure used the probabilities assigned at stage 2 above.

5 There were links between the choices of days of the week in different segments of the year and also

between the choices of specific dates. These links were designed to rule out certain combinations of days for the reasons described above. However they did not affect the overall chance of selection of any particular day, as the next two paragraphs explain.

The procedures for assigning particular days of the week to each segment gave each day of the week the same probability in all the linked segments. The chance of assigning any particular day of the week to a given segment was thus the same as it would have been if the selections had been made separately with the same probabilities.

The procedures for linking the selection of specific dates in different segments all had the following characteristics. A segment was divided into two or more portions and each date is assigned to one of these portions, or partially assigned to two of them. The linking procedure decided which portion was to contain the interviewing day. For each specific date the chance of choosing the relevant portion (or portions) times the probability, having done so, of picking the specific date was equal to the probability assigned to that date at stage 2. These characteristics of the linking procedures meant that each date's chance of selection was equal to the probability it was originally assigned.

6 Days which were expected to be busy were given a higher than average chance of selection. One way of doing this was to assign the days concerned comparatively high probabilities of selection at stage 2. This was mainly done for half-term and bank holidays at the Science Museum and the National Railway Museum.

Two steps were taken at all the museums to increase the selection chances of busy weekend days:

a. The procedure for stage 3 gave Sundays, or rather Sunday afternoons since the museums do not open until 2.30 on Sunday, enhanced representation to allow for the very high hourly attendance.

b. The segments into which the year was divided at stage 1 for the purpose of selecting weekend days were of unequal length—being shorter in the summer than in winter to give busy summer weekends a higher chance of selection.

Selecting weekdays: details
Ten of the 16 days in each museum's sample were weekdays. For the purpose of selecting weekdays the year was divided into the following ten segments.

A	Christmas holidays	20 December 1979—6 January 1980
B	Spring term, first ½	7 January—23 February 1980
C	Spring term, second ½	24 February—1 April 1980
D	Easter holidays	2 April—20 April 1980
E	Summer term, first ½	21 April—9 June 1980
F	Summer term, second ½	10 June—22 July 1980
G	Summer holidays, first ½	23 July—11 August* 1980
H	Summer holidays, second ½	11 August*—2 September 1980
I	Autumn term, first ½	3 September—4 November 1980
J	Autumn term, second ½	{ 5 November—2 December 1980
		{ 3—19 December 1979

---

* 11 August was assigned partly to the first half and partly to the second half of the summer holidays.

Probabilities were then assigned to the specific dates in segments A and D. In the remaining segments the dates falling on different days of the week were listed separately. Probabilities were then assigned to each date representing its chance of selection given that the choice of day fell on that particular day of the week. The probabilities were assigned separately at each museum in order to take account of known differences in attendance patterns.

The next stage was to assign a day of the week to each segment (except segments A and D). At the Science Museum and the National Railway Museum the procedure was as follows. One of the five days of the week was selected for segment B—each day having an equal chance of selection. The choice for segment C was made from the four remaining days—the choice again being made with equal probabilities. The same procedure was followed in segment E with the remaining three days, and in segment F with the remaining two days. The one remaining weekday was then assigned to segment I. The same procedure was then used to allocate days of the week to segments J, G and H.

The procedure at the Victoria and Albert Museum was basically the same. However since the museum was closed on Fridays there were only four days of the week to assign. The procedure was therefore run through twice: first to assign days of the week to segments B, C, E and F and then to assign days to segments I, J, G and H.

The next step was to choose what portions of segments B, C, E, F, I and J the selected dates should fall into. Each of these segments were divided into a first portion, a middle portion, and a final portion—the dates in each having a third of the initially assigned probabilities for that segment (particular dates could be assigned partly to one portion and partly to the following portion). One of the three portions in segment B was then chosen—each having been given equal probability. A choice was then made for segment E. If the first third of segment B had been chosen the choice for segment E would be between the second and third portions each of which was given equal probability, and similarly if the second or third portions of segment B had been chosen. The remaining third was allocated to segment I.

The same third was assigned to segment J as to segment B. The procedure used to allocate thirds to segments F and C was the same as that used for segments E and I.

The final step was to choose specific dates in each segment with probabilities proportional to those originally assigned.

Selecting weekend days: details
Six of the 16 days at each museum were allocated to weekend days—Saturdays, Sundays or bank holidays. The survey year was divided into the following six unequal segments:

K   8 November—30 November 1980
       8 December 1979—6 January 1980
L   12 January—23 March 1980
M   29 March—11 May 1980
N   17 May—20 July 1980
O   26 July—31 August 1980
P   6 September—2 November 1980

At each museum probabilities were assigned to the specific dates in every segment. This was done separately for Saturdays on the one hand and for Sundays/ bank holidays on the other. At all three museums the probability that the date selected for segment K would be in the holiday period between 22 December and 5 January (Saturdays) or 23 December and 6 January (Sundays) was set at 40 per cent. The chance that the date selected for segment M would be in the holiday period between 5 April and 19 April (Saturdays) or 6 April and 20 April (Sundays/bank holidays) was set at 60 per cent.

The museums were open for three bank holidays during the survey year. These were Easter Monday (7 April), the Spring Bank Holiday (Monday 26 May) and the Summer Bank Holiday (Monday 25 August). For the purposes of the selection procedure Easter Day was also treated as a bank holiday. These bank holidays fell in segments M, N and O of the survey year. For each of these three segments two sets of probabilities were drawn up—one for the procedure that could select a bank holiday and the other for the procedure that could not select one. In the first set the bank holidays were assigned twice their initial chance of selection and a reduced probability was assigned to each of the Sundays. In the second set the bank holidays were given no chance of selection and the probability of selection of each Sunday was enhanced. These probabilities were calculated so that the probability initially assigned to each date was the average (arithmetic mean) of the probabilities under the two different procedures.

At each museum two of the segments were assigned for Saturday interviews. The choice was made with equal probability between the following three pairs of segments (K, N), (L, O) and (M, P). As a result one of the three bank holiday segments (M, N and O) would have been allocated for Saturday interviewing. The next step was to assign one of the two remaining bank holiday segments to the procedure that could select a bank holiday and the other to the procedure that could not. This was done with equal probability.

Once these steps had been completed it was possible to select specific dates. The selections for segments N, O and P were made independently. However the selections for segments K, L and M were linked, segments K and M contain the Christmas and Easter holidays. If the selection for segment M fell in the holiday period the day chosen for segment K did not, and vice versa. Segment K was also linked with segment L. If the date selected for segment K came from the first half of the segment, so did the date selected for segment L.

Calculating each date's chance of selection

The probability that a particular date would be chosen for the sample was equal to

> [the probability that the relevant day of the week was selected for that segment] × [the probability originally assigned to the particular date, given that the selection was to be made from dates on that day of the week].

The value of the first term is $\frac{1}{2}$ for Saturdays, $\frac{2}{3}$ for Sundays or bank holidays, $\frac{1}{5}$ for weekdays at the Science and National Railway Museums, and $\frac{1}{4}$ for weekdays at the Victoria and Albert Museum.

In the case of the Christmas and Easter holiday weekday segments there was no initial selection of the day of the week. For these two segments the chance of picking a given day was simply the probability originally assigned to the date in question.

Choice of counting intervals

The basic operating principle of the count-based samples was that interviewers should attempt to speak to every 'n'th person leaving the particular museum on the day in question. The choice of the interval 'n' was made in the light of the expected attendance at the museum that day and of the number of interviews that the interviewing team could carry out in the time. Each day's counting interval was set before the day itself but, except for the first few days, after the start of the survey year. It was thus possible to use the experience of the first part of the survey year when setting the counting intervals for subsequent days.

Weighting the data

All the count-based tables in the main body of this report use weighted data. The weighting factors used took account of:

> The individual visitor's (strictly the individual *visit's*) chance of selection

> The information available about non-response.

i. Allowing for the individual visitor's chance of selection

The initial probability that an individual visitor would be included in the sample was equal to:

> [the original probability that the date in question would be picked] ÷ [the counting interval on the day concerned].

The weighting factor applied to the information provided by each visitor was proportional to:

> 1 ÷ [the visitor's initial probability of being included in the sample].

Tables A.1, A.2 and A.3 show the probabilities of selection for the 16 days included in the sample at each of the museums. The final column of the tables shows the weighting factor applied for individuals interviewed on each day.

Table A.1 Selection probabilities and weighting factors for days in the count-based sample at the Victoria and Albert Museum

(a) Segment	(b) Day selected	(c) Probability of selecting day of week	(d) Conditional probability of selecting specific date given (c)	(e) Probability of selecting specific date $(=c \times d)$	(f) Counting interval	(g) Probability of selecting individual visitor $\left(=\dfrac{c \times d}{f}\right)$	(h) Weighting factor $\left(=\dfrac{1}{100} \times \dfrac{1}{g}\right)$
A	Weds 2 January	—	1/5	1/5	40	1/200	2.00
B	Thurs 17 January	1/4	43/300	1/27.91	40	1/1116	11.16
C	Mon 10 March	1/4	1/6	1/24	40	1/960	9.60
D	Thurs 17 April	—	11/100	1/9.09	60	1/545	5.45
E	Tues 3 June	1/4	43/300	1/27.91	50	1/1395	13.95
F	Weds 16 July	1/4	1/6	1/24	60	1/1440	14.40
G	Weds 23 July	1/4	33/100	1/12.12	60	1/727	7.27
H	Thurs 14 August	1/4	33/100	1/12.12	60	1/727	7.27
I	Mon 29 September	1/4	11/100	1/36.36	40	1/1455	14.55
J	Tues 11 November	1/4	43/300	1/27.91	30	1/837	8.37
K	Sun 16 November	2/3	12/125	1/15.63	80	1/1250	12.50
L	Sat 12 January	1/3	9/100	1/33.33	40	1/1333	13.33
M	Sun 6 April	2/3	3/20	1/10	150	1/1500	15.00
N	Sun 8 June	2/3	9/100	1/16.67	100	1/1667	16.67
O	Sat 16 August	1/3	17/100	1/17.65	60	1/1059	10.59
P	Sun 5 October	2/3	3/25	1/12.5	80	1/1000	10.00

Table A.2 Selection probabilities and weighting factors for days in the count-based sample at the Science Museum

(a) Segment	(b) Day selected	(c) Probability of selecting day of week	(d) Conditional probability of selecting specific date given (c)	(e) Probability of selecting specific date $(=c \times d)$	(f) Counting interval	(g) Probability of selecting individual visitor $\left(=\dfrac{c \times d}{f}\right)$	(h) Weighting factor $\left(=\dfrac{1}{100} \times \dfrac{1}{g}\right)$
A	Weds 2 January	—	13/100	1/7.69	100	1/769	7.69
B	Tues 5 February	1/5	19/150	1/39.47	40	1/1579	15.79
C	Fri 28 March	1/5	1/5	1/25	100	1/2500	25.00
D	Thurs 10 April	—	9/100	1/11.11	150	1/1667	16.67
E	Mon 9 June	1/5	1/6	1/30	50	1/1500	15.00
F	Thurs 19 June	1/5	1/6	1/30	100	1/3000	30.00
G	Weds 6 August	1/5	33/100	1/15.15	120	1/1818	18.18
H	Fri 29 August	1/5	33/100	1/15.15	60	1/909	9.09
I	Wed 10 September	1/5	1/10	1/50	50	1/2500	25.00
J	Tues 2 December	1/5	43/300	1/34.88	40	1/1395	13.95
K	Sun 23 December	2/3	33/250	1/11.36	200	1/2273	22.73
L	Sun 16 March	2/3	9/100	1/16.67	300	1/5000	50.00
M	Sat 26 April	1/3	1/10	1/30	100	1/3000	30.00
N	Sun 1 June	2/3	9/100	1/16.67	100	1/1667	16.67
O	Sun 13 August	2/3	7/50	1/10.71	120	1/1286	12.86
P	Sat 27 September	1/3	11/100	1/27.27	70	1/1909	19.09

Table A.3 Selection probabilities and weighting factors for days in the count-based sample at the National Railway Museum

(a) Segment	(b) Day selected	(c) Probability of selecting day of week	(d) Conditional probability of selecting specific date given (c)	(e) Probability of selecting specific date $(=c \times d)$	(f) Counting interval	(g) Probability of selecting individual visitor $\left(=\dfrac{c \times d}{f}\right)$	(h) Weighting factor $\left(=\dfrac{1}{100} \times \dfrac{1}{g}\right)$
A	Thurs 3 January	—	13/100	1/7.69	20	1/154	1.54
B	Tues 8 January	1/5	19/150	1/39.47	10	1/395	3.95
C	Thurs 20 March	1/5	1/5	1/25	15	1/375	3.75
D	Thurs 17 April	—	2/25	1/12.5	30	1/375	3.75
E	Mon 19 May	1/5	1/6	1/30	20	1/600	6.00
F	Weds 9 July	1/5	1/6	1/30	60	1/1800	18.00
G	Weds 23 July	1/5	33/100	1/15.15	100	1/1515	15.15
H	Tues 26 August	1/5	17/50	1/14.71	100	1/1471	14.71
I	Fri 24 October	1/5	1/10	1/50	15	1/750	7.50
J	Fri 7 November	1/5	1/6	1/30	10	1/300	3.00
K	Sat 29 December	1/3	17/125	1/22.06	30	1/662	6.62
L	Sun 2 March	2/3	9/100	1/16.67	50	1/833	8.33
M	Sun 11 May	2/3	1/10	1/15	75	1/1125	11.25
N	Sat 7 June	1/3	1/10	1/30	40	1/1200	12.00
O	Sun 31 August	2/3	3/25	1/12.5	200	1/2500	25.00
P	Sun 19 October	2/3	11/100	1/13.64	60	1/818	8.18

ii. Reweighting for non-response

Table A.4 shows the extent of non-response from various sources at each museum. To provide some information about the effect of refusals and non-contacts interviewers were asked to record the sex and apparent age of anyone who they failed to interview and to note whether the individual concerned seemed to have been visiting the museum alone, in an informal group or in an organised party. The results showed that non-contacts and refusals were particularly high among 6–10 year olds visiting the museums in organised parties, especially girls. To correct for this it was decided to reweight the answers of 6–10 year olds so that the total weight attached to the answers of visitors in each of the four categories (6–10 year old boys in organised parties; *ditto* not in organised parties; 6–10 year old girls in organised parties; *ditto* not in organised parties) was equal to what it would have been if the weighted response rate in each of the categories had been equal to the average weighted response rate among non-organised party 6–10 year olds in the museum concerned. For this purpose the response rate was defined as the proportion of children—other than those who turned back into the museum—who agreed to be interviewed.

The reweighting factors for 6–10 year olds are listed in Table A.5. The weight applied to the answers of each 6–10 year old in the report tables is equal to:

[the weight needed to correct for varying probabilities of selection] × [the reweighting factor to allow for non-response among 6–10 year olds].

Non-response is discussed further below in the section on the reliability of the survey's results.

iii. Calculating the weighted base numbers for the count-based tables

The total of the weights calculated for all the eligible individuals who were successfully interviewed as part of the count-based sample at each museum, came to several times the actual unweighted sample size. It was thought that the figures would be more easily interpreted by readers if the weighted base number was close to the actual number of individuals contributing to it. As a result it was decided to set the weighted base for the whole eligible co-operating count-based sample at 1,000 for each museum. Where figures are based on part of the sample, the weighted base numbers are proportional to the total weight assigned to the individuals in that part of the sample. The weighted base numbers have been rounded to the nearest ten.

The quota samples

The quota sample interviews with people leaving each museum were spread between different days of the week and different times of year. However the days were not picked on a probability basis. Instead they were assigned with an eye to the convenience of organising the interviewing work. For this reason runs of several consecutive days were generally selected, days were chosen so as not to clash with count-based interviewing days, and no quota interviewing took place during the Christmas and Easter holidays. The quota interviews with people leaving specific galleries were all conducted during the first three months of the survey year.

No quota interviews were sought with children aged ten or less since it was thought that they would not be able to deal with the questions in the quota-interview questionnaires. Nor were visitors with organised parties asked for interviews since it was not thought possible to separate them from their groups for long enough to conduct one of the quota interviews. This consideration applied particularly to visitors leaving the museum. In the case of the interviews with people leaving specific galleries children visiting with school parties were included in the set quotas.

Each interviewer was given a quota of interviews to complete—12 on weekdays and Saturdays and six on Sundays. The quotas of people leaving the museums were generally set in terms of sex, age and whether the individual was visiting the museum alone or with others. For instance the quota might specify one interview with a man aged 21–30 visiting the museum alone, two interviews with women aged 31 or over visiting the museum in company, and so on. On Sundays the quotas simply specified interviews with one man and one woman in each of three age groups without stipulating whether or not they were accompanied.

Table A.4 Response and non-response of the count-based sample at each museum

	Victoria and Albert Museum	Science Museum	National Railway Museum
	%	%	%
Not contact			
Interviewing finished	5	9	3
Other	1	2	7
Refusal			
Could not speak English	3	3	0
No time	7	9	5
Other	2	1	1
Total refusals and non-contacts	18	24	16
Ineligible	8	8	1
Turned back into museum	1	0	0
Staff on delivery	5	5	0
Not first visit of the day	3	3	1
Interview	74	67	83
Base	*1,210*	*1,166*	*1,118*

The percentages in this table are derived from weighted data—weighted to allow for differing probabilities of selection but not for the effects of non-response. The base numbers are the unweighted total of individuals selected by the count.

Table A.5 Factors used to reweight for non-response by 6–10 year old children

Children aged 6–10	Victoria and Albert Museum	Science Museum	National Railway Museum
Girls in organised parties	1.241	1.854	1.508
Girls not in organised parties	0.957	1.026	1.051
Boys in organised parties	1.536	1.156	1.241
Boys not in organised parties	1.052	0.989	0.934

Table A.6 Composition of set quotas

Type of quota	Set number	Sex	Age		Visiting:
Weekday and	2	Males	11–20		With others
Saturday quota	1	Males	21–30	(21–40 V & A)	Alone
(Type 1)	1	Males	21–30	(21–40 V & A)	With others
	1	Males	31 +	(41 + V & A)	Alone
	1	Males	31 +	(41 + V & A)	With others
	1	Females	11–20		Alone
	1	Females	11–20		With others
	1	Females	21–30	(21–40 V & A)	Alone
	1	Females	21–30	(21–40 V & A)	With others
	2	Females	31 +	(41 + V & A)	With others
Weekday and	1	Males	11–20		Alone
Saturday quota	1	Males	11–20		With others
(Type 2)	2	Males	21–30	(21–40 V & A)	With others
	1	Males	31 +	(41 + V & A)	Alone
	1	Males	31 +	(41 + V & A)	With others
	1	Females	11–20		Alone
	1	Females	11–20		With others
	2	Females	21–30	(21–40 V & A)	With others
	1	Females	31 +	(41 + V & A)	Alone
	1	Females	31 +	(41 + V & A)	With others
Weekday and	1	Males	11–20		Alone
Saturday quota	1	Males	11–20		With others
(Type 3)	1	Males	21–30	(21–40 V & A)	Alone
	1	Males	21–30	(21–40 V & A)	With others
	2	Males	31 +	(41 + V & A)	With others
	2	Females	11–20		With others
	1	Females	21–30	(21–40 V & A)	Alone
	1	Females	21–30	(21–40 V & A)	With others
	1	Females	31 +	(41 + V & A)	Alone
	1	Females	31 +	(41 + V & A)	With others
Sunday quota	1	Males	11–20		—
	1	Males	21–30	(21–40 V & A)	—
	1	Males	31 +	(41 + V & A)	—
	1	Females	11–20		—
	1	Females	21–30	(21–40 V & A)	—
	1	Females	31 +	(41 + V & A)	—

different types of weekday quota were set at each museum, Table A.6 gives details. Each of the quotas contains two men and two women in each of three age groups, four people (two men and two women) visiting the museum alone and eight people (four men and four women) visiting the museum with other people. Three interviewers worked at the exit of each London museum, one working with each type of quota. Only two interviewers worked at the exit of the National Railway Museum—there the quotas were rotated so that every three days two quotas of each type were issued.

The quotas set targets for the interviewers but were hardly ever completely fulfilled. Tables A.7 to A.12 describe the samples of people with whom interviews were actually achieved (and give comparable information for the respondents in the count-based samples). It can be seen that the quotas have acted as a kind of corset, constraining but not determining the shape of the achieved samples. Thus Table A.7 shows that, despite men and women being given equal numbers of places in the set quotas the quota sample of people leaving the Science Museum contains more males than females. Nevertheless the preponderance of men and boys is less than it would be if the quota samples had simply reflected the composition of the museum's visiting public as shown in the figures from the count-based sample. Table A.8 looks at the proportions of informants who were visiting the museum alone or with family or friends. In all three museums the quota samples, as intended, contain a higher pro-

portion of solitary visitors than the count-based samples. Even so, the lower proportion of lone visitors at the two technical museums than at the Victoria and Albert Museum—demonstrated by the count-based samples—is reflected to some extent in the composition of the quota samples.

Different age bands were used to set the quotas at the Science Museum and the Victoria and Albert Museum since it was believed, correctly, that the Science Museum's visiting public was younger than that of the Victoria and Albert Museum. It was thought that the National Railway Museum's visitors would resemble the Science Museum's public in this respect, and so the same age bands were used as for the Science Museum. However, as Table A.7 shows, the National Railway Museum has as high a proportion of older visitors as the Victoria and Albert Museum. The National Railway Museum's older visitors are therefore comparatively under-represented in the quota samples.

The quotas set for interviewers working at individual galleries inside the museums were the same as those shown in Table A.6, except that on Mondays to Fridays the quotas included two interviews with 11–20 year olds (one boy and one girl) visiting the museums with school parties. These quotas each included two fewer interviews with 11–20 year olds visiting the museums with others but not in an organised group. (Chapter 11 discusses the results of including school parties in the gallery samples.) Since only one interviewer worked at each gallery during the gallery interviewing days

Table A.7 Comparison of count-based and quota samples: sex and age

	Victoria and Albert Museum			Science Museum			National Railway Museum		
	Count-based sample	Quota sample (exit)	Quota sample (galleries)	Count-based sample	Quota sample (exit)	Quota sample (galleries)	Count-based sample	Quota sample (exit)	Quota sample (galleries)
	%	%	%	%	%	%	%	%	%
Sex									
Male	43	46	49	64	56	61	61	54	57
Female	57	54	51	36	44	39	39	46	43
Age (years)									
0–10	6	—	—	19	—	—	11	—	—
11–20	23	25	27	36	32	34	21	31	27
21–30	28	26	24	15	31	31	15	33	36
31–40	16	13	14	15	16	16	27	14	16
41 and over	27	35	35	14	21	19	26	22	20
Base	*891*	*1,003*	*959*	*787*	*1,045*	*947*	*920*	*596*	*511*
Weighted base	*1,000*	*—*	*—*	*1,000*	*—*	*—*	*1,000*	*—*	*—*

Table A.8 Comparison of count-based and quota samples: whether alone or in a group

	Victoria and Albert Museum			Science Museum			National Railway Museum		
	Count-based sample	Quota sample (exit)	Quota sample (galleries)	Count-based sample	Quota sample (exit)	Quota sample (galleries)	Count-based sample	Quota sample (exit)	Quota sample (galleries)
	%	%	%	%	%	%	%	%	%
Alone	26	36	34	10	27	21	5	21	20
With friends but not members of family	24	28	26	20	31	26	11	26	29
With members of family	38	36	34	45	42	40	64	53	44
With an organised party	12	—	6	25	—	12	20	—	6
Base	*891*	*1,003*	*959*	*787*	*1,045*	*947*	*920*	*596*	*511*
Weighted base	*1,000*	*—*	*—*	*1,000*	*—*	*—*	*1,000*	*—*	*—*

(except for the Main Hall at the National Railway Museum where two interviewers were stationed) the different types of weekday quota were rotated.

Reliability of the survey's results

The reliability of any survey's results depends on the representativeness of the sample of people interviewed and on the extent of sampling variability—the random variation that causes the estimates derived from any sample to differ slightly from the results that would be obtained if it were practicable to interview everybody in the population being surveyed.

A previous section of this appendix explained how the count-based samples of museum leavers were selected. After reweighting the figures to allow for different selection probabilities the results from the selected sample would be fully representative if everyone selected actually took part in the survey. However in practice not everyone selected for the sample could be interviewed. Table A.4 set out the reasons for this non-response.

Twenty-six per cent of (weighted) selections at the Victoria and Albert Museum, 33 per cent of selections at the Science Museum and 17 per cent of selections at the National Railway Museum failed to produce an eligible interview. However not all this non-response represents missing data. A considerable proportion, at the two London museums particularly, consisted of people who were ineligible either because they were not visitors at all but staff or people making deliveries to

the museum, or because they changed their mind and did not leave the museum after all, or because when they were interviewed they said that they had already left the museum earlier in the day and therefore had already had their chance of selection. A better measure of the proportion of interviews lost by non-response is therefore the percentage of non-contacts and refusals—18 per cent at the Victoria and Albert Museum, 24 per cent at the Science Museum and 16 per cent at the National Railway Museum.

These figures themselves give only an approximate idea of the proportion of eligible interviews lost by non-response since we do not know what proportion of the refusals and non-contacts may themselves have been ineligible. It is however possible to calculate an upper limit for the proportion of eligible interviews lost as a result of non-response by assuming that all the non-contacts and refusals would have been eligible and percentaging them on the total of eligible interviews plus non-contacts and refusals. This gives an upper-limit estimate of the loss from non-response of 20 per cent for the Victoria and Albert Museum, 27 per cent for the Science Museum and 16 per cent for the National Railway Museum.

At all three museums interviewing had to stop some time before the museum doors actually closed in order to allow the museum staff to clear the museums by six o'clock. This was the main reason for non-contacts at the two London museums. The main reason for refusals at all three museums was shortage of time. However at both London museums several people were counted as refusals because their English was not good enough for even the short count-based interview.

The question of representativeness takes a different guise in the case of the quota samples. Since the purpose of the quota interviews was to estimate relationships between visitors' characteristics and their patterns of visiting and reactions to what they saw, the object in setting the quotas was to obtain adequate numbers of interviews with visitors in the different categories defined by age, sex and whether they were visiting alone or with other people. As pointed out above, this meant that the distributions of various characteristics among people interviewed as part of the quota samples differ from their distributions among each museum's visiting public as a whole. To this extent the quotas were intentionally unrepresentative. However quotas can also produce samples that are biased in ways that were not intended. This can happen because the people who are willing to give up 20 minutes or so to talk to an interviewer may differ in various ways from those who are not. Tables A.9 to A.12 examine various areas in which bias might arise.

Table A.9 compares the length of time visitors in the quota samples of museum leavers had spent in the museums with the length of visits reported by members of the count-based samples. The two sets of distributions turn out to be very similar. At the two London museums the quota samples contain a smaller proportion of people reporting visits of less than an hour.

However this may owe more to differences in the questions used to obtain the data than to real differences between the samples. (The fact that people who left the museum more than once on the interviewing day were only eligible for the count-based sample on the first occasion would also have led to a slight difference between the distributions in the two types of sample.)

Table A.10 compares the reasons given for visiting each museum by members of the two kinds of sample. Fewer members of the quota samples have visited the museums in order to accompany someone else, which is what one would expect given that the quotas deliberately over-represented solitary visitors. The quota samples contain more people who are visiting the museum out of general interest or because of the museum's reputation and considerably more casual or holiday visitors than the count-based samples. However the proportions attracted by special exhibitions or citing their interest in the museum's contents as a main reason for their visits are very much the same with either type of sample.

Table A.11 relates to a factor (who, if anyone, was accompanying the informant) which was used to set the target quotas. It shows that members of the quota samples who were accompanied by other people tended to be accompanied by fewer people than members of the count-based samples. The explanation may be that individuals in large groups who might have been willing to be interviewed themselves did not want to keep the

Table A.9 Comparison of count-based and quota samples: time spent in museums

Time spent in museum	Victoria and Albert Museum		Science Museum		National Railway Museum	
	Count-based sample	Quota sample (exit)	Count-based sample	Quota sample (exit)	Count-based sample	Quota sample (exit)
	%	%	%	%	%	%
Less than 1 hour	31	22	22	16	29	30
1 hour but less than 2 hours	43	47	41	42	53	52
2 hours but less than 3 hours	18	21	25	30	14	16
3 hours or more	8	9	12	12	3	3
Base	891	1,003	787	1,045	920	596
Weighted base	1,000	—	1,000	—	1,000	—

Table A.10 Comparison of count-based and quota samples: reasons given for visiting each museum

Reasons for visiting museum	Victoria and Albert Museum		Science Museum		National Railway Museum	
	Count-based sample	Quota sample (exit)	Count-based sample	Quota sample (exit)	Count-based sample	Quota sample (exit)
	%	%	%	%	%	%
To accompany someone else	11	9	33	23	40	24
General interest, reputation of the museum	21	29	18	25	19	31
Because the visitor liked the museum	5	9	8	11	7	8
Interest in the museum's contents	25	25	27	25	37	38
To see a special temporary exhibition	26	26	10	11	0	0
In connection with work or studies	7	3	5	5	3	2
To use information facilities	5	2	1	0	0	1
Casual or holiday visit	9	18	13	23	21	38
Other	12	5	8	6	4	4
Base	891	1,003	787	1,045	920	596
Weighted base	1,000	—	1,000	—	1,000	—

The percentages add to over 100% as people could give more than one answer.

Table A.11 Comparison of count-based and quota samples: number of companions

Companions	Victoria and Albert Museum		Science Museum		National Railway Museum	
	Count-based sample	Quota sample (exit)	Count-based sample	Quota sample (exit)	Count-based sample	Quota sample (exit)
	%	%	%	%	%	%
Alone	30	36	13	28	6	21
With one other person	40	51	30	48	26	45
With two other people	12	8	20	12	16	12
With three or more people	18	5	37	13	52	21
Base	*802*	*1,003*	*620*	*1,045*	*755*	*596*
Weighted base	*880*	*—*	*750*	*—*	*800*	*—*

other members of the group waiting for the 20 minutes or so needed for an interview. Table A.12 compares the age distribution of members of the 11–20 age group in the count-based and quota samples. The quota samples appear to be somewhat biased towards the upper end of the age group.

So far this section has dealt with the representativeness of each type of sample. It is now time to consider the effect of the random variation involved in sample selection. Sampling variability affects the results from both the count-based and quota-based samples. The problem for the reader is how to decide which estimates are reliable and which are less certain. Formal confidence intervals and tests of significance have not been used in this report. However the text of the report does provide guidance. The more reliable results have been reported in definite language while those which depend on more variable figures have been described

in a tentative way. The more reliable results are those based on larger numbers. Where a contrast between different museums is drawn it is more reliable, other things being equal, if the percentage difference in the factor of interest is large. Relationships between visitors' characteristics and their viewing patterns and reactions are most reliable if they are based on comparable figures from several different galleries or from all three museums, or if the results derived from the gallery interviews are corroborated by data from the samples of museum leavers.

Some readers will want to draw their own conclusions from the tables presented in this report—highlighting aspects not discussed in the text. When doing so it is important to take account of the factors discussed in the preceding paragraphs, and particularly the points in the last paragraph.

Table A.12 Comparison of count-based and quota samples: age distribution of 11–20 year olds

Age (years)	Victoria and Albert Museum			Science Museum			National Railway Museum		
	Count-based sample	Quota sample (exit)	Quota sample (galleries)	Count-based sample	Quota sample (exit)	Quota sample (galleries)	Count-based sample	Quota sample (exit)	Quota sample (galleries)
	%	%	%	%	%	%	%	%	%
11	4	3	2	12	3	4	14	6	14
12	7	5	4	12	2	8	15	7	9
13	7	4	6	12	7	14	14	12	9
14	10	8	9	13	9	13	15	7	10
15	9	6	9	16	11	12	16	9	10
16	13	8	7	9	13	11	8	10	9
17	9	8	13	9	10	11	5	6	6
18	10	17	12	7	14	8	3	13	10
19	14	15	16	4	18	9	3	13	11
20	17	25	23	6	14	9	6	16	12
Base	*206*	*255*	*261*	*290*	*337*	*322*	*191*	*187*	*140*
Weighted base	*230*	*—*	*—*	*360*	*—*	*—*	*210*	*—*	*—*

Appendix 2   The schedules

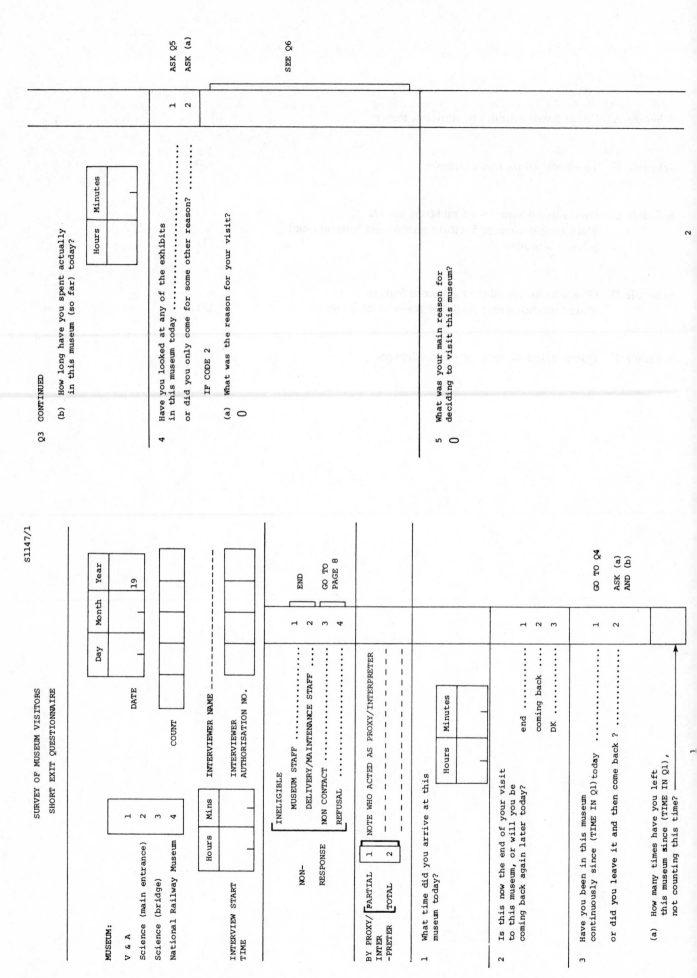

S1147/1

SURVEY OF MUSEUM VISITORS
SHORT EXIT QUESTIONNAIRE

MUSEUM:

V & A	1
Science (main entrance)	2
Science (bridge)	3
National Railway Museum	4

DATE

Day	Month	Year
		19

COUNT

INTERVIEWER NAME _ _ _ _ _ _ _ _ _

INTERVIEWER
AUTHORISATION NO.

INTERVIEW START TIME

Hours	Mins

INELIGIBLE		
	MUSEUM STAFF	1	END
NON-	DELIVERY/MAINTENANCE STAFF	2	
RESPONSE	NON CONTACT	3	GO TO
	REFUSAL	4	PAGE 8

BY PROXY/	PARTIAL	1	NOTE WHO ACTED AS PROXY/INTERPRETER
INTER -PRETER	TOTAL	2	_ _ _ _ _ _ _ _ _

1 What time did you arrive at this
museum today?

Hours	Minutes

2 Is this now the end of your visit
to this museum, or will you be
coming back again later today?

end	1
coming back	2
DK	3

3 Have you been in this museum
continuously since (TIME IN Q1) today 1   GO TO Q4

or did you leave it and then come back ? 2   ASK (a)
AND (b)

(a) How many times have you left
this museum since (TIME IN Q1),
not counting this time? _ _ _ _ _ _ _ _

1

Q3 CONTINUED

(b) How long have you spent actually
in this museum (so far) today?

Hours	Minutes

4 Have you looked at any of the exhibits
in this museum today   1   ASK Q5

or did you only come for some other reason?   2   ASK (a)

IF CODE 2

(a) What was the reason for your visit?

0

} SEE Q6

5 What was your main reason for
deciding to visit this museum?

0

2

116

Page 3

NATIONAL RAILWAY MUSEUM ONLY

DNA AT SCIENCE MUSEUM AND V&A ...X → ASK Q7

6 Is there any particular reason why you chose this time of year for your visit?
Yes ..... 1 → ASK (a)
No ...... 2 → ASK Q7

IF YES
(a) What was that? → ASK Q7

TO ALL

7 Apart from this museum can you tell me roughly how many museums and art galleries you have visited in the last 12 months?
None 1
1 2
2 to 4 3
5 to 9 4
10 or more ... 5
DK 6

8 Is today the first time you have visited this museum?
Yes ... 1 → ASK Q9
No .... 2 → ASK (a)

IF NO
(a) How many times have you been before?
1 1
2 to 4 2
5 to 9 3
10 or more 4
DK 5

9 Did you come to this museum today as part of a school party, or a coach tour, or any other sort of organised party?
Yes ... 1 → ASK Q10
No .... 2 → GO TO Q11

3

Page 4

IF WITH ORGANISED PARTY

10 What sort of organised party did you come with?
UK school party ..... 1 → ASK (a)
Other (SPECIFY, STATING WHETHER INFORMANT GUIDE, ORDINARY MEMBER OF THE GROUP, OR OTHER) ..... 2 → GO TO Q15

IF UK SCHOOL PARTY
(a) Are you ..... AS APPROPRIATE [a teacher ..... 1 → GO TO Q15
a pupil ..... 2 → GO TO Q15
or do you have some other connection with the visit (SPECIFY)? 3

IF NOT WITH ORGANISED PARTY

11 Did you come to this museum today
on your own .... 1 → GO TO Q15
with friends .... 2 → SEE Q12
with members of your family 3
or with anyone else? (SPECIFY) 4

ASK IF WITH MEMBERS OF FAMILY
[DNA OTHERWISE ... X → GO TO Q13

12 Which members of your family came here with you today?
CODE ALL THAT APPLY
spouse .... 1 → ASK (a)
son 2
daughter 3
father 4
mother 5
brother 6
sister 7
other (SPECIFY) 8

(a) COMPLETE FOR EACH OF INFORMANT'S CHILDREN AT MUSEUM

CHILD NO	1	2	3	4	5	6	7
son	1	1	1	1	1	1	1
daughter	2	2	2	2	2	2	2
AGE LAST BIRTHDAY							

4

117

13 Altogether, how many people are visiting this
    museum with you today (not counting yourself)? ──────▶

14 May I just check, how many of them are aged

5 or less?	
6 to 10?	
INDIVIDUAL	11 to 15? ....
PROMPT	16 to 20? ....
	21 to 30? ....
	31 to 50? ....
	51 to 70? ....
	71 or more? ....

BACKGROUND

15 What was your age on your last
    birthday? ──────▶

16 SEX

| MALE ..... | 1 |
| FEMALE ... | 2 |

17 Do you live in the United Kingdom
    or somewhere else?

| UK .... | 1 | ASK Q18 |
| Somewhere else .... | 2 | GO TO Q19 |

IF UK

18 What town or village and what
    county do you live in?

    Town ....
    (Borough if in GLC)

    County ....   SEE Q21

IF LIVES OUTSIDE UK

19 What country do you live in?

    DNA IRELAND, AUSTRALIA
    NEW ZEALAND, USA X   SEE Q21

OTHER RESIDENTS ABROAD

20 May I just check, what is your first language?

EDUCATION

    DNA 15 OR UNDER X   END
    DNA 50 OR OVER ..... Y   ASK Q22

21 Have you completed your
    full-time education?

| Yes .... | 1 | ASK Q22 |
| No ..... | 2 | ASK Q23 |

IF FULL-TIME EDUCATION COMPLETED

22 At what age did you finish your
    full-time education?

    DNA FINISHED AGED
    16 OR UNDER X   ASK Q24

IF FINISHED AGED 17 OR OVER

(a) What were the main subjects you studied
    during your last two years of full-time
    education?   ASK Q24

IF FULL-TIME EDUCATION NOT COMPLETED

23 What are the main subjects you have studied
    during the last two years?   END

5

FULL-TIME EDUCATION COMPLETED

24 At present are you

CODE FIRST
THAT APPLIES

working full-time	1	
working part-time	2	
retired	3	ASK Q25
unemployed and looking for work	4	
a housewife	5	
or not working for some other reason? (SPECIFY)	6	END

25 OCCUPATION

IF 1,2 AT Q24 PRESENT OCCUPATION

IF 3 LAST OCCUPATION PRIOR TO RETIREMENT

IF 4 LAST OCCUPATION

INTERVIEWER CHECKS

S/E	1
Employee	2
Manager	1
Foreman/Sup ..	2
other	3

NOTE: IF TEACHER/LECTURER PROBE FOR SUBJECTS TEACHES AND TYPE OF SCHOOL/COLLEGE TEACHES AT

END

7

NON-RESPONSE SECTION

A EXPLAIN REASONS FOR REFUSAL/NON-CONTACT

DESCRIPTION OF INFORMANT

B SEX

MALE	1
FEMALE	2

C ESTIMATE AGE

5 or less	1
6 to 10	2
11 to 15	3
16 to 20	4
21 to 30	5
31 to 50	6
51 to 70	7
71 or more	8

D DID INDIVIDUAL APPEAR TO BE IN AN ORGANISED PARTY?

Yes	1	END
No	2	GO TO E

E IF NO TO D
DID INDIVIDUAL APPEAR TO BE VISITING THE MUSEUM ON HIS OWN?

ALONE	1
WITH OTHERS	2

Social Survey Division
Office of Population Censuses & Surveys
St Catherines House
10 Kingsway
LONDON WC2B 6JP

W847 OPCS 9/79

8

119

Schedule B To obtain quota-based samples

S1147/2

SURVEY OF MUSEUM VISITORS
QUOTA QUESTION FORM

	Day	Month	Year
DATE			19

MUSEUM:
V & A 1
Science 2
National Railway Museum ... 4

EXIT QUOTA .... 1 → QUOTA NUMBER
GALLERY QUOTA ...... 2 → GALLERY NAME

QUOTA NUMBER

GALLERY NAME

LETTER

INTERVIEWER NAME – – – – – – –

INTERVIEWER
AUTHORISATION NO.

INTERVIEW NUMBER

INELIGIBLE
MUSEUM STAFF 1
DELIVERY/MAINTENANCE STAFF ... 2   END
DOES NOT SPEAK ENGLISH 3

1 Did you come to this museum today as
   part of a school party, or a coach tour,
   or any other sort of organised party?
   Yes 1   ASK Q2
   No 2   GO TO Q3

2 What sort of organised party did you
   come with?
   UK school party ... 1   ASK (a)
   Other (SPECIFY) ... 2   END

   DNA EXIT QUOTAS ... X

IF UK SCHOOL PARTY (GALLERIES ONLY)
(a) Are you .....   AS   [a teacher 1   END
                    APPROPRIATE [a pupil 2   SEE Q4
                    or do you have some other
                    connection with the visit? 3   END

1

IF NOT WITH SCHOOL PARTY

3 Did you come to this museum today
   CODE   on your own 1
   ALL   with friends 2   SEE Q4
   THAT   with members of your family 3
   APPLY   or with anyone else? (SPECIFY).... 4

   DNA GALLERY QUOTAS ... X

EXIT QUOTAS ONLY
4 Have you looked at any of the exhibits
   in this museum today 1   ASK Q5
   or did you only come for some other reason? . 2   ASK Q6

   DNA GALLERY QUOTAS ... X
                                                     1   ASK Q6
                                                     2   END

GALLERY QUOTAS ONLY
5 You are just leaving [PARTICULAR GALLERY]
   May I just check, did you stop to look at
   any of the things in [PARTICULAR GALLERY] ... 1   ASK Q6
   or did you walk straight through without
   stopping to look at the exhibits? 2   END

TO ALL
6 Is this now the end of your visit to
   [MUSEUM/PARTICULAR GALLERY] or will
   you be coming back again later today?
                              End of visit ... 1   ASK Q7
                              Coming back 2   END
                              DK 3   ASK Q7

7 What was your age on your last birthday?

8 SEX   Male 1
                              Female 2

9 May I just check, have you already
   been interviewed in connection with
   this survey today?
                              Yes 1   END
                              No 2   SEE INTERVIEWER CHECK

INTERVIEWER CHECK
IS INFORMANT ELIGIBLE?
                              Yes 1   START MAIN INTERVIEW
                              No 2   END

Social Survey Division
Office of Population Censuses & Surveys
St Catherines House
10 Kingsway

WR46 OPCS 9/79

Schedule C Quota-based samples of museum leavers
Main questionnaire: Victoria and Albert Museum and Science Museum

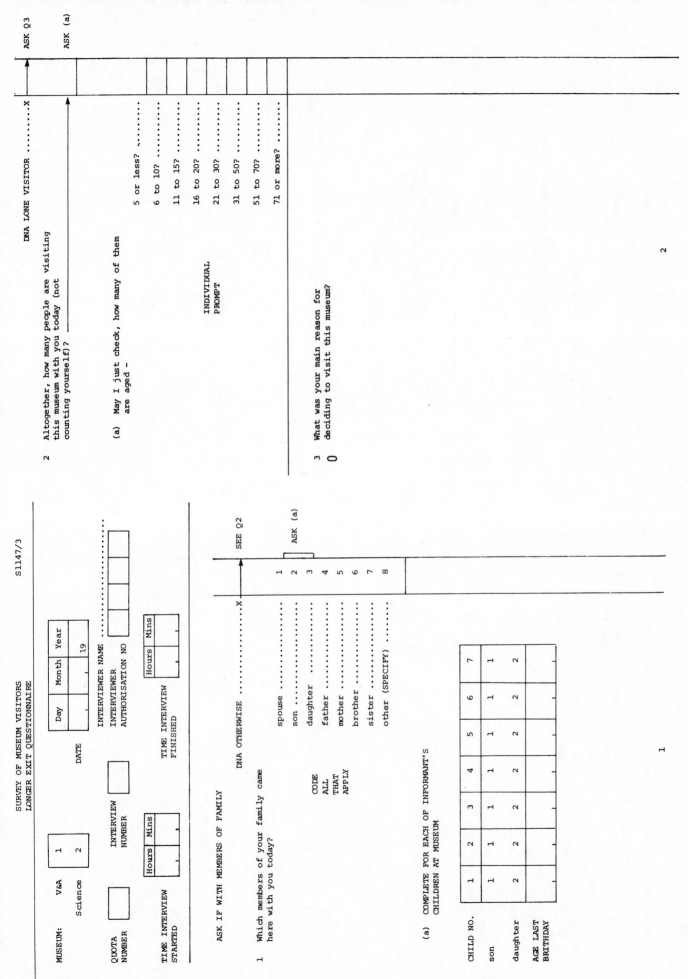

S1147/3

SURVEY OF MUSEUM VISITORS
LONGER EXIT QUESTIONNAIRE

MUSEUM: V&A 1
         Science 2

DATE   Day Month Year
                            19

INTERVIEWER NAME
INTERVIEWER AUTHORISATION NO

QUOTA NUMBER

INTERVIEW NUMBER

TIME INTERVIEW STARTED   Hours Mins
TIME INTERVIEW FINISHED   Hours Mins

ASK IF WITH MEMBERS OF FAMILY

1   Which members of your family came here with you today?

DNA OTHERWISE X   → SEE Q2

CODE ALL THAT APPLY
spouse 1
son 2
daughter 3   → ASK (a)
father 4
mother 5
brother 6
sister 7
other (SPECIFY) 8

(a) COMPLETE FOR EACH OF INFORMANT'S CHILDREN AT MUSEUM

CHILD NO.	1	2	3	4	5	6	7
son	1	1	1	1	1	1	1
daughter	2	2	2	2	2	2	2
AGE LAST BIRTHDAY							

1

ASK Q3

ASK (a)

DNA LONE VISITOR X   → ASK Q3

2   Altogether, how many people are visiting this museum with you today (not counting yourself)?

(a) May I just check, how many of them are aged –

INDIVIDUAL PROMPT
5 or less? ....
6 to 10? ....
11 to 15? ....
16 to 20? ....
21 to 30? ....
31 to 50? ....
51 to 70? ....
71 or more? ....

3   What was your main reason for deciding to visit this museum?
0

2

4 Did you decide to come out of general interest...... 1 | ASK Q5

RUNNING PROMPT

or will this visit to the museum help you with your job . 2

your studies 3

or with a definite hobby? ...... 4 | ASK (a)

CODE ALL THAT APPLY

OTHER (SPECIFY) ...... 5 | ASK Q5

ASK SEPARATELY FOR EACH OF 2/3/4 RINGED

(a) How will this visit to the museum help you with your job/studies/hobby?

5 Has your visit to this museum today been

RUNNING PROMPT

very enjoyable 1

fairly enjoyable 2

or not really enjoyable? 3

6 Some people find that going round a museum can be tiring. Now that you are leaving do you feel

RUNNING PROMPT

very tired 1

fairly tired 2

or not really tired? 3

Now I would like to ask about what you have done during your visit to this museum today.

7 How long have you spent in this museum today?

HOURS	MINS

8 Here is a plan of the museum.
I would like you to tell me which
parts you have looked at today, in
the order you went round.

In which section did you first
stop and look at something?
THEN REPEAT UNTIL THIS INTERVIEW
REACHED
In which section did you next
stop and look at something?
NOTE ANSWERS IN FIRST COLUMN

9 Now I would like you to show me on
the plan how you got from one
section to the next.

How did you get to (FIRST
SECTION)?
THEN REPEAT UNTIL THIS
INTERVIEW REACHED
How did you get to (NEXT
SECTION)?
RECORD ANSWERS ON PLAN.
BE SURE TO SHOW HOW INFORMANT
REACHED EXIT.

ORDER OF VISIT	SECTION NUMBER GIVEN ON CHART	Very interesting	Fairly interesting	Not really interesting	Doesn't remember clearly	OFFICE USE ONLY
1st		1	2	3	4	
2nd		1	2	3	4	
3rd		1	2	3	4	
4th		1	2	3	4	
5th		1	2	3	4	
6th		1	2	3	4	
7th		1	2	3	4	
8th		1	2	3	4	
9th		1	2	3	4	
10th		1	2	3	4	
11th		1	2	3	4	
12th		1	2	3	4	
13th		1	2	3	4	
14th		1	2	3	4	
15th		1	2	3	4	

		Very interesting	Fairly interesting	Not really interesting	Doesn't remember clearly	OFFICE USE ONLY
16th		1	2	3	4	
17th		1	2	3	4	
18th		1	2	3	4	
19th		1	2	3	4	
20th		1	2	3	4	
21st		1	2	3	4	
22nd		1	2	3	4	
23rd		1	2	3	4	
24th		1	2	3	4	
25th		1	2	3	4	
26th		1	2	3	4	
27th		1	2	3	4	
28th		1	2	3	4	
29th		1	2	3	4	
30th		1	2	3	4	

10 Now I would like to ask how
interesting or uninteresting you
personally found each part of
the museum.

What do you feel about
(FIRST SECTION).

Was it -

RUNNING   very interesting
PROMPT   fairly interesting
          or not really interesting?

REPEAT FOR EACH SECTION IN ORDER OF VISIT

123

DNA MORE THAN ONE
SECTION LOOKED AT .....X → ASK Q12

11 IF ONLY ONE SECTION LOOKED AT
Had you decided before arriving at the museum that you would look at (SECTION LOOKED AT) today?

Yes 1
No 2
ASK Q13

12 IF MORE THAN ONE SECTION LOOKED AT
Altogether which section did you personally find the most interesting?

SECTION NUMBER ..
DK .. 99

13 Had you (or the people you have come with) decided before arriving at the museum today that there was anything you definitely wanted to see on this visit?

Yes ..... 1 ASK (a)
No ..... 2 SEE Q14

IF YES
(a) What did you (or they) definitely want to see?

NB ANSWERS CAN INCLUDE OBJECTS/ SECTIONS NOT ACTUALLY SEEN BY INFORMANT.
PROBE FULLY FOR DETAILS

8

DNA NOT VISITING WITH OWN
CHILD AGED 10 OR LESS .....X → GO TO Q15

14 TO PARENT VISITING WITH OWN CHILDREN AGED 10 OR LESS
Which section of this museum do you think your children found most interesting during their visit today?

SECTION NUMBER
DK 99
NB SECTION NEED NOT HAVE BEEN LOOKED AT BY INFORMANT

15 During your visit to this museum today did you ask any member of staff for directions, or for any other information?

Yes .. 1 ASK (a)-(b)
No ... 2 GO TO Q16

IF YES
(a) Did you ask them

INDIVIDUAL PROMPT   to help you find your way round?.. 1
CODE ALL THAT APPLY for information about a particular exhibit?................. 2
                     for any other kind of information? (SPECIFY) 3

(b) Did you find their answers
0 RUNNING PROMPT   very helpful 1
                    fairly helpful 2
                    or not really helpful?.. 3
GO TO Q16
ASK (i)

IF CODE 3
(i) Why weren't their answers really
0 helpful?

9

16 May I just check, did you visit the
    museum shop today?

Yes ..	1	ASK (a)
No ...	2	GO TO Q17

IF YES
(a) Did you buy anything there?

Yes ..	1	ASK (i)
No ...	2	GO TO (b)

(i) Did you buy

	Yes	No	
any postcards?	1	2	
INDIVIDUAL PROMPT any books?.........	1	2	
any posters?	1	2	
CODE 1 OR 2 anything else?	1	2	
FOR EACH (SPECIFY)			

(b) Do you feel that improvements could
    be made in what the shop sells, or in
    the way it is set out and run?

Yes ..	1	ASK (i)
No ...	2	ASK Q17

IF YES
(i) In what ways do you feel
    improvements could be made?
O

10

17 May I just check, did you visit the

museum [tea room / restaurant] (SCIENCE) (V&A) today?

Yes ..	1	ASK (a)
No ...	2	SEE Q18

IF YES
(a) Do you feel that improvements could be
    made in the way the [tea room / restaurant] is set out
    and run, or in the quality of the food?

Yes ..	1	ASK (i)
No ...	2	SEE Q18

IF YES
(i) In what ways do you feel
    improvements could be made?
O

11

125

DNA SCIENCE MUSEUM ....X → ASK Q19

18 V&A ONLY
During your visit today did you buy

	Yes	No
INDIVIDUAL PROMPT — a plan of this museum?	1	2
a guide-book for this museum?	1	2
CODE ALL THAT APPLY — a guide or a study-sheet for part of this museum?	1	2

ASK Q20

19 SCIENCE MUSEUM ONLY
During your visit today did you -

	Yes	No
buy a plan of this museum?	1	2
pick up or buy any other kind of guide to all or part of this museum? (SPECIFY)	1	2

20 We would like to know how clear the
0   information provided in the entrance hall is.
Can you think back to the beginning of your
visit today, when you were in the entrance
hall and were deciding which part of this
museum to visit first.

	Yes	No
Did you realise then that there was a plan of the museum in the entrance hall?	1	2
Did you realise then that you could buy a plan of this museum to take round with you?	1	2
And did you realise then that there were lifts (SCIENCE: and escalators) to the upper floors?	1	2

21 Before you came to this museum today had
0   you any idea that the museum

INDIVIDUAL PROMPT — puts on lectures which are open to the public?	1	2
CODE 1 OR 2 FOR EACH — has a library which is open to the public?	1	2
has expert staff who give opinions on objects the public bring in?	1	2

22 Do you feel that improvements <u>need</u> to be made in
0

	Yes	No	DK
1. the seating in the museum?	1	2	3
2. the museum's heating and air-conditioning system?	1	2	3
3. the lighting in the museum?	1	2	3
4. the signs provided to help people find their way round the museum?	1	2	3
5. the museum's opening and closing times?	1	2	3
6. any other feature of the museum? (SPECIFY)	1	2	3

INDIVIDUAL PROMPT

FOR EACH ASPECT CODED 1, RECORD ASPECT NO. AND ASK
(a) In what way do you feel improvements need to be made? (PROBE FULLY)
0

ASPECT NUMBER _____

ASPECT NUMBER _____

ASPECT NUMBER _____

ASPECT NUMBER _____

14

Q22 continued

ASPECT NUMBER _____

ASPECT NUMBER _____

23 Apart from this museum can you tell me roughly how many museums and art galleries you have visited in the last 12 months?

None	1
1	2
2 to 4	3
5 to 9	4
10 or more	5
DK	6

24 Different museums cover different subjects. Do you personally find some subjects which museums cover more interesting than others?
0

Yes ..	1	ASK (a)
No ...	2	GO TO Q25

IF YES
(a) What subjects are they?
0

15

25 Is today the first time you have visited this museum?

Yes ..	1	ASK (a)
No ...	2	GO TO (c)

IF YES
(a) How does the content of the museum compare with what you expected.

RUNNING PROMPT

Is it more or less what you expected .	1	GO TO (b)
or is it different?	2	ASK (a1)
DK	3	GO TO (b)

IF CODE 2 AT (a)
(a1) How does the content differ from what you expected?

0

(b) Has today's visit made you want to come back and see more of this museum ....

RUNNING PROMPT

....	1	GO TO Q26
or do you feel that you have now seen as much as you want to?	2	

IF CODE 2 AT MAIN Q
(c) How many times have you been before?

1	1
2 to 4	2
5 to 9	3
10 or more ...	4
DK	5

(d) Apart from this visit how many times have you been in the last 12 months?

None	1
1	2
2 to 4	3
5 to 9	4
10 or more ..	5
DK	6

26 People visit museums for all kinds of reasons and we have listed a few of them here
(SHOW CARD 1)

0

When you visit a museum - any museum - which reason usually applies to you <u>most</u>?

I go because it is something to do ...	1
I go to keep someone else company	2
I go for my own interest	3
I go because I have been told to	4

27 When you have been round a museum is it important to you to feel that you have learnt something

0

....	1
or do you just enjoy looking at things?	2

28 When you look round a room in a museum do you like to think about the different objects individually

0

....	1
or do you prefer them to tell an overall story?	2

INFORMANT'S BACKGROUND

29 Do you live in the United Kingdom or somewhere else?

UK 1   SEE Q30
somewhere else ... 2   ASK (a)

IF SOMEWHERE ELSE
(a) What country do you live in?

DNA IRELAND, AUSTRALIA
NEW ZEALAND, USA X   SEE Q30

OTHER RESIDENTS ABROAD
(b) May I just check, what is your first language?

DNA 15 OR UNDER X   END
DNA 50 OR OVER ...... Y   ASK Q31

30 ALL AGED 16-49
Have you completed your full-time education?

Yes .. 1   ASK Q31
No ... 2   GO TO Q32

31 IF FULL-TIME EDUCATION COMPLETED
At what age did you finish your full-time education?

DNA FINISHED AGED 16 OR UNDER X   ASK Q33

IF FINISHED AGED 17 OR OVER
(a) What were the main subjects you studied during your last two years of full-time education?

ASK Q33

32 IF FULL-TIME EDUCATION NOT COMPLETED
What are the main subjects you have studied during the last two years?

END

130

33 FULL-TIME EDUCATION COMPLETED

At present are you

CODE FIRST THAT APPLIES

working full-time 1
working part-time .... 2 ⎱ ASK Q34
retired 3
unemployed and looking for work 4
a housewife 5
or not working for some other reason? (SPECIFY) .... 6 ⎱ END

34 OCCUPATION

IF 1, 2 AT Q33 PRESENT OCCUPATION

IF 3 LAST OCCUPATION PRIOR TO RETIREMENT

IF 4 LAST OCCUPATION

NOTE: IF TEACHER/LECTURER
PROBE FOR SUBJECTS TEACHES
AND TYPE OF SCHOOL/COLLEGE
TEACHES AT

INTERVIEWER CHECKS

S/E 1
Employee 2

Manager 1
Foreman/Sup 2
Other 3

Social Survey Division
Office of Population Censuses and Surveys
St Catherines House
10 Kingsway
LONDON WC2B 6JP

W848 OPCS 9/79

END

20

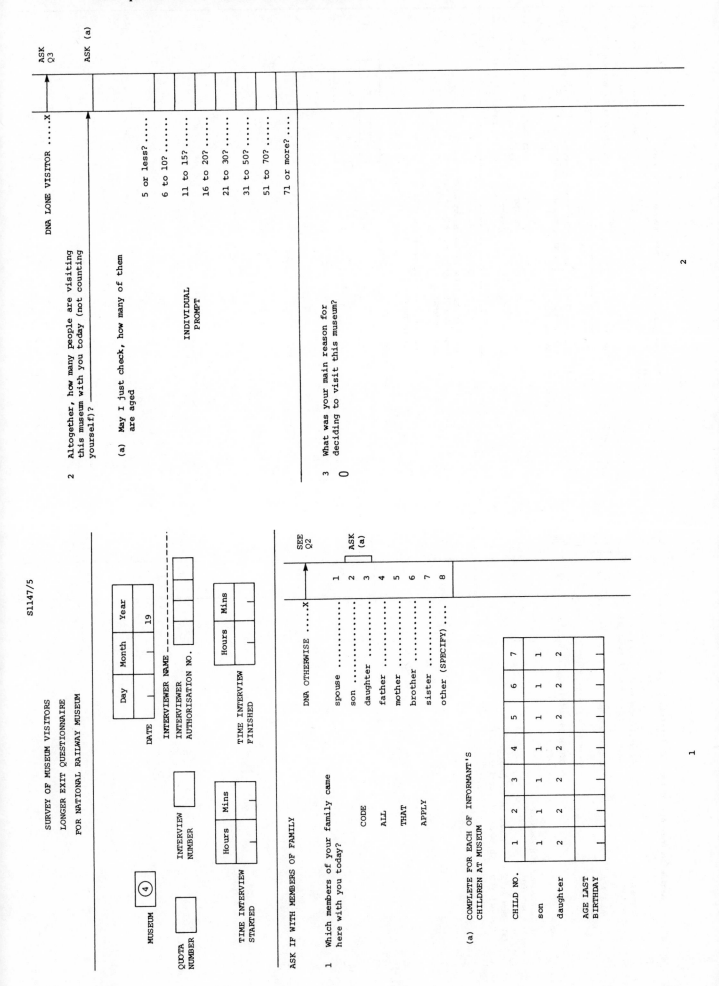

S1147/5

SURVEY OF MUSEUM VISITORS
LONGER EXIT QUESTIONNAIRE
FOR NATIONAL RAILWAY MUSEUM

MUSEUM [4]

INTERVIEW NUMBER [ ]

QUOTA NUMBER [ ]

DATE | Day | Month | Year 19

INTERVIEWER NAME _____

INTERVIEWER AUTHORISATION NO.

TIME INTERVIEW STARTED | Hours | Mins

TIME INTERVIEW FINISHED | Hours | Mins

ASK IF WITH MEMBERS OF FAMILY

1 Which members of your family came here with you today?

CODE ALL THAT APPLY

DNA OTHERWISE X

spouse 1
son 2   ASK (a)
daughter 3
father 4
mother 5
brother 6
sister 7
other (SPECIFY) ... 8

SEE Q2

(a) COMPLETE FOR EACH OF INFORMANT'S CHILDREN AT MUSEUM

CHILD NO.	1	2	3	4	5	6	7
son	1	1	1	1	1	1	1
daughter	2	2	2	2	2	2	2
AGE LAST BIRTHDAY	_	_	_	_	_	_	_

2 Altogether, how many people are visiting this museum with you today (not counting yourself)?

DNA LONE VISITOR X   ASK Q3

(a) May I just check, how many of them are aged

INDIVIDUAL PROMPT

5 or less? ......
6 to 10? ......
11 to 15? ......
16 to 20? ......
21 to 30? ......
31 to 50? ......
51 to 70? ......
71 or more? ....

ASK (a)

3 What was your main reason for deciding to visit this museum?

0

1

2

4 Did you decide to come out of general interest 1   ASK Q5

    RUNNING
    PROMPT

or will this visit to the museum help you with your job... 2

    CODE ALL
    THAT
    APPLY

your studies 3

or with a definite hobby? 4   ASK (a)

OTHER (SPECIFY)............ 5   ASK Q5

ASK SEPARATELY FOR EACH OF 2/3/4 RINGED

(a) How will this visit to the museum help you with your job/studies/hobby?

3

---

5 Has your visit to this museum today been

    RUNNING
    PROMPT

very enjoyable 1

fairly enjoyable 2

or not really enjoyable? 3

6 Some people find that going round a museum can be tiring. Now that you are leaving do you feel

    RUNNING
    PROMPT

very tired 1

fairly tired 2

or not really tired? 3

Now I would like to ask about what you have done during your visit to this museum today.

7 How long have you spent in this museum today?

Hours	Mins

4

DNA ONLY
ONE SECTION LOOKED AT

IF MORE THAN ONE SECTION LOOKED AT

10   Altogether which section did you personally find the most interesting?

SECTION NUMBER....	
ONE SECTION LOOKED AT..X	→ ASK Q11
DK.... 99	

11   Had you (or the people you have come with) decided before arriving at the museum today that there was anything you definitely wanted to see on this visit?

Yes.... 1   ASK (a)
No..... 2   SEE Q12

IF YES

(a) What did you (or they) definitely want to see?

NB. ANSWERS CAN INCLUDE OBJECTS/SECTIONS NOT ACTUALLY SEEN BY INFORMANT. PROBE FULLY FOR DETAILS

---

8   Here is a plan of the museum. I would like you to tell me which parts of it you looked at today.

	Yes	No	Very interesting	Fairly interesting	Not really interesting	Doesn't remember clearly
Starting with the main hall, did you look at the locomotives on turntable A here (POINT)?	1	2	1	2	3	4
And did you look at the coaches on turntable B here (POINT)?	1	2	1	2	3	4
And did you stop and look at any of the exhibits round the side of the main hall?	1	2	1	2	3	4
Did you look at the exhibition in the room here (POINT) under the balcony?	1	2	1	2	3	4
Did you watch any of the slide show in the tape and slide theatre in the balcony over the main hall?	1	2	1	2	3	4
Did you look at any of the other things in the balcony over the main hall?	1	2	1	2	3	4
Did you look at anything in the front gallery to the right of the entrance here (POINT)?	1	2	1	2	3	4

9
0   Now I would like to ask how interesting or uninteresting you personally found each of the parts of the museum which you looked at today.

ASK IN ORDER FOR SECTIONS CODED 1 (YES)

What do you feel about [SECTION]. Was it

RUNNING   very interesting
PROMPT   fairly interesting
             or not really interesting?

DNA NOT VISITING WITH OWN CHILD AGED 10 OR LESS....X   GO TO Q13

TO PARENT VISITING WITH OWN CHILDREN AGED 10 OR LESS

12   Which section of this museum do you think your children found most interesting during their visit today?

SECTION NUMBER .. |__|__|

DK .. 99

NB. SECTION NEED NOT HAVE BEEN LOOKED AT BY INFORMANT

13   During your visit to this museum today did you ask any member of staff for directions, or for any other information?

Yes ...	1	ASK (a)-(b)
No	2	GO TO Q14

IF YES

(a)   Did you ask them

INDIVIDUAL PROMPT

CODE ALL THAT APPLY

to help you find your way round? ...	1
for information about a particular exhibit?	2
for any other kind of information? .	3

(SPECIFY)

(b)   Did you find their answers

RUNNING PROMPT

very helpful	1	GO TO Q14
fairly helpful	2	GO TO Q14
or not really helpful? ...	3	ASK (i)

IF CODE 3

(i)   Why weren't their answers really helpful?

0

14   May I just check, did you visit the museum shop today?

Yes ...	1	ASK (a)
No	2	GO TO Q15

IF YES

(a)   Did you buy anything there?

Yes ...	1	ASK (i)
No	2	GO TO (b)

IF YES

(i)   Did you buy

INDIVIDUAL PROMPT

CODE 1 OR 2 FOR EACH

	Yes	No
any postcards? ...	1	2
any books?	1	2
any posters?	1	2
anything else? ...	1	2

(SPECIFY)

(b)   Do you feel that improvements could be made in what the shop sells, or in the way it is set out and run?

0

Yes ...	1	ASK (i)
No	2	ASK Q15

IF YES

(i)   In what ways do you feel improvements could be made?

0

15

15 May I just check, did you visit the museum refreshment room today?

Yes .... 1 ASK (a)
No .... 2 GO TO Q16

IF YES

(a) Do you feel that improvements could be made in the way the refreshment room is set out and run, or in the quality of the food?

0

Yes ... 1 ASK (i)
No .... 2 GO TO Q16

IF YES

(i) In what ways do you feel improvements could be made?

0

16

16 During your visit today did you buy

		Yes	No
INDIVIDUAL	a guide book for this museum? ...	1	2
PROMPT	a guide or a study-sheet for a part of this museum?	1	2

9

17

17 Do you feel that improvements need to be made in

0

	Yes	No	DK
1 the seating in the museum?	1	2	3
2 the museum's heating and air conditioning system?	1	2	3
3 the lighting in the museum?	1	2	3
4 the signs provided to help people find their way round the museum?	1	2	3
5 the museum's opening and closing times?	1	2	3
6 any other feature of the museum? (SPECIFY)	1	2	3

FOR EACH ASPECT CODED 1, RECORD ASPECT NO. AND ASK

(a) In what way do you feel improvements need to be made? [PROBE FULLY]

0

ASPECT NUMBER _____

ASPECT NUMBER _____

ASPECT NUMBER _____

ASPECT NUMBER _____

10

Q17 CONTINUED

ASPECT NUMBER ————→

ASPECT NUMBER ————→

18 Apart from this museum can you tell me roughly how many museums and art galleries you have visited in the last 12 months?

None	1	
1	2	
2 to 4	3	
5 to 9	4	
10 or more	5	
DK	6	

19 Different museums cover different subjects. Do you personally find some subjects which museums cover more interesting than others?

Yes	1	ASK (a)
No	2	GO TO Q20

IF YES

(a) What subjects are they?

---

20 Is today the first time you have visited this museum?

Yes	1	ASK (a)
No	2	GO TO (c)

IF YES

(a) How does the content of the museum compare with what you expected.

RUNNING PROMPT

Is it more or less what you expected?	1	GO TO (b)
or is it different?	2	ASK (a1)
DK	3	GO TO (b)

IF CODE 2 AT (a)

(a1) How does the content differ from what you expected?

(b) Has today's visit made you want to come back and see more of this museum

RUNNING PROMPT

	1	GO
or do you feel that you have now seen as much as you want to?	2	TO Q21

IF CODE 2 AT MAIN Q

(c) How many times have you been before?

1	1
2 to 4	2
5 to 9	3
10 or more	4
DK	5

INFORMANT'S BACKGROUND

24 Do you live in the United Kingdom or somewhere else?

UK ..	1	SEE Q25
somewhere else ..	2	ASK (a)

IF SOMEWHERE ELSE

(a) What country do you live in?

DNA IRELAND, AUSTRALIA NEW ZEALAND, USA	X	SEE Q25

OTHER RESIDENTS ABROAD

(b) May I just check, what is your first language?

DNA 15 OR UNDER	X	END
DNA 50 OR OVER	Y	ASK Q26

25 Have you completed your full-time education?

Yes....	1	ASK Q26
No.....	2	GO TO Q27

(d) Apart from this visit how many times have you been in the last 12 months?

None	1
1	2
2 to 4	3
5 to 9	4
10 or more	5
DK	6

21 People visit museums for all kinds of reasons and we have listed a few of them here. [SHOW CARD 1]

When you visit a museum - any museum - which reason usually applies to you most?

I go because it is something to do	1
I go to keep someone else company	2
I go for my own interest	3
I go because I have been told to	4

22 When you have been round a museum is it important to you to feel that you have learnt something

	1
or do you just enjoy looking at things?	2

23 When you look round a room in a museum do you like to think about the different objects individually ......

	1
or do you prefer them to tell an overall story?	2

IF FULL TIME EDUCATION COMPLETED

26 At what age did you finish your full-time education

   DNA FINISHED AGED 16 OR UNDER ..... X → ASK Q28

IF FINISHED AGED 17 OR OVER

(a) What were the main subjects you studied during your last two years of full-time education? → ASK Q28

IF FULL-TIME EDUCATION NOT COMPLETED

27 What are the main subjects you have studied during the last two years? → END

FULL-TIME EDUCATION COMPLETED

28 At present are you

   CODE FIRST THAT APPLIES

   working full-time .... 1 ⎫ ASK Q29
   working part-time .... 2 ⎬
   retired 3   ⎭
   unemployed and looking for work 4
   a housewife 5   ⎫ END
   or not working for some other reason? (SPECIFY) .... 6 ⎭

29 OCCUPATION

IF 1,2 AT Q28 PRESENT OCCUPATION
IF 3 LAST OCCUPATION PRIOR TO RETIREMENT
IF 4 LAST OCCUPATION

NOTE: IF TEACHER/LECTURER PROBE FOR SUBJECTS TEACHES AND TYPE OF SCHOOL/COLLEGE WHERE TEACHES

INTERVIEWER CHECKS

   S/E 1
   Employee 2

   Manager 1
   Foreman/Sup.. 2
   Other 3

END

Social Survey Division
Office of Population Censuses & Surveys
St Catherines House
10 Kingsway
LONDON WC2B 6JP

W850 OPCS 9/79

Schedule E Quota-based samples of gallery leavers

S1147/4

SURVEY OF MUSEUM VISITORS

GALLERY QUESTIONNAIRE

MUSEUM:

V&A	1
Science	2
National Railway Museum	4

DATE

Day	Month	Year
		19

GALLERY LETTER [ ]

INTERVIEW NUMBER

INTERVIEWER NAME ....

INTERVIEWER AUTHORISATION NO [ ][ ][ ][ ]

TIME INTERVIEW STARTED

Hours	Mins

TIME INTERVIEW FINISHED

Hours	Mins

IF WITH MEMBERS OF FAMILY

DNA OTHERWISE .... X → GO TO Q2

1 Which members of your family have come to this museum with you today?

CODE ALL THAT APPLY

spouse	1
son	2
daughter	3
father	4
mother	5
brother	6
sister	7
Other (SPECIFY)	8

son / daughter → ASK (a)-(b) PAGE 2

1

(a) COMPLETE FOR EACH OF INFORMANT'S CHILDREN AT MUSEUM

CHILD NO	1	2	3	4	5	6	7
son	1	1	1	1	1	1	1
daughter	2	2	2	2	2	2	2
AGE LAST BIRTHDAY	.	.	.	.	.	.	.
Yes	1	1	1	1	1	1	1
No	2	2	2	2	2	2	2
DNA	3	3	3	3	3	3	3

ASK FOR EACH CHILD AGED 10 OR UNDER

(b) Has .... just been visiting (PARTICULAR GALLERY) with you?

2

139

2 Here is a plan showing (PARTICULAR GALLERY). Can you show me what you stopped to look at during your visit today? ENTER ITEM NUMBERS IN BOX BELOW

(a) Can we just look over the rest of the plan.

Did you stop and look at anything (else) over here?
INDICATE ABOUT A QUARTER OF THE PLAN

REPEAT SEPARATELY FOR EACH OF THE OTHER THREE QUARTERS
RECORD IN BOX FURTHER ITEMS INDICATED BY INFORMANT

ITEM NUMBERS									

(b) Does that cover everything you stopped to look at in (PARTICULAR GALLERY) today,

or did you stop and look at anything else which you can't place on the plan?

1	GO TO Q3
2	ASK (i)

IF CODE 2
(i) What else did you stop and look at?

3 Is today the first time you have visited (PARTICULAR GALLERY*) in this museum?
(* IN MAIN HALL NRM 'this museum')

Yes ..	1	ASK Q4
No ...	2	ASK (a)

IF NO
(a) How many times before have you visited (PARTICULAR GALLERY*)?
(* IN MAIN HALL NRM 'this museum')

1	1
2 to 4	2
5 to 9	3
10 or more ..	4
DK	5

4 About how long have you spent in (PARTICULAR GALLERY) today?

Hours	Mins

5 0 Did you find (PARTICULAR GALLERY)

RUNNING
PROMPT

very interesting	1
fairly interesting	2
or not really interesting?	3

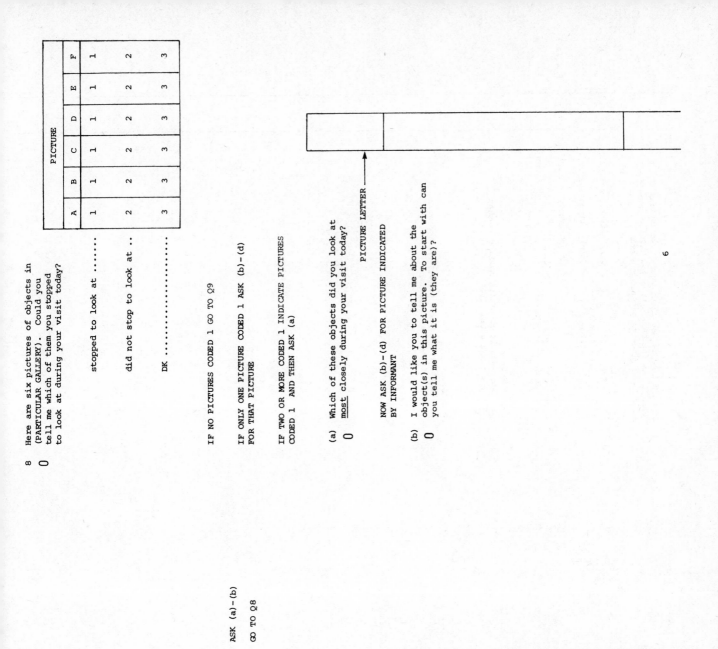

8   Here are six pictures of objects in (PARTICULAR GALLERY). Could you tell me which of them you stopped to look at during your visit today?

	PICTURE					
	A	B	C	D	E	F
stopped to look at	1	1	1	1	1	1
did not stop to look at ..	2	2	2	2	2	2
DK	3	3	3	3	3	3

IF NO PICTURES CODED 1 GO TO Q9

IF ONLY ONE PICTURE CODED 1 ASK (b)-(d) FOR THAT PICTURE

IF TWO OR MORE CODED 1 INDICATE PICTURES CODED 1 AND THEN ASK (a)

(a) Which of these objects did you look at most closely during your visit today?

PICTURE LETTER

NOW ASK (b)-(d) FOR PICTURE INDICATED BY INFORMANT

(b) I would like you to tell me about the object(s) in this picture. To start with can you tell me what it is (they are)?

6

---

6   Did you find (PARTICULAR GALLERY)

very enjoyable	1
RUNNING fairly enjoyable	2
PROMPT or not really enjoyable?	3

7   Was there one object in (PARTICULAR GALLERY) which you found especially interesting?

Yes ..	1	ASK (a)-(b)
No ...	2	GO TO Q8

IF YES
(a) Which object was that?

(b) May I just check, where is (OBJECT) on the plan?

RECORD ITEM NUMBER

DK ...   99

5

I would like to ask about your first impression of (PARTICULAR GALLERY). When you first walked into it did you find its general appearance ....

RUNNING PROMPT	very attractive	1 ASK (a1)
	fairly attractive	2
	or not really attractive?	3 ASK (a2)

9
0

IF CODE 1 OR 2
(a1) What was it about its appearance that you found attractive?
0

IF CODE 3
(a2) What was it about its appearance that you found unattractive?
0

(c) What do you think are the main points of interest about it (them)?
0

(d) Is there anything else you can tell me about it (them)?
0

7

11 Do you think there is anyway the general presentation of (PARTICULAR GALLERY) could be improved?

Yes .. 1   ASK (a)

No ... 2   ⎱ SEE Q12

DK ... 3   ⎰

IF YES
(a) In what ways do you think it could be improved?

10

---

10 Do you think that all the things in (PARTICULAR GALLERY) are displayed equally well............ 1 GO TO Q11

or are some things better displayed than others?........ 2 ASK (a)-(c)

DK .... 3 GO TO Q11

IF CODE 2
(a) Which things do you think are best displayed?

(b) Why do you think that (THINGS) are well displayed?

(c) May I just check, where are (THINGS) on the plan? NOTE ITEM NUMBERS IN BOX

ITEM NUMBERS

9

DNA FOR GALLERIES A,B,C,D,G,K ———————→ GO TO Q13

FOR GALLERIES E,F,H,J

12 Now I would like to ask you about the information provided in (PARTICULAR GALLERY)

HAND CARD 2

On this card we have listed the different ways information is provided in (PARTICULAR GALLERY). Could you tell me whether you have

	Yes	No	
read any of the labels or other written information ?...............	1	2	
listened to any of the telephone recordings?..	1	2	
INDIVIDUAL PROMPT listened to any of the loudspeaker recordings?............	1	2	
ASK ONLY FOR SOURCES LISTED looked at any of the television films?............	1	2	
ON CARD looked at any of the slide shows?.........	1	2	
looked at the slide show in the tape and slide theatre?............	1	2	
looked at any of the small slide displays?.........	1	2	

IF NO INFORMATION SOURCE USED RING X ————————→ SEE Q17

DNA 2 OR MORE SOURCES USED..X ————————→ ASK (b)

IF ONLY ONE SOURCE USED

(a) Why did you prefer just to (APPROPRIATE PHRASE FROM MAIN QUESTION)?

0

GO TO Q14

IF TWO OR MORE SOURCES USED

(b) Which of these kinds of information, that is (SOURCES USED), did you prefer overall?

0

HAND CARD 2

written	1	
telephone	2	
loudspeaker	3	
television	4	ASK (i)
slide shows	5	
tape & slide theatre .	6	
small slide displays .	7	
DK	8	GO TO Q14

(i) Why do you say that?

0

GO TO Q14

FOR GALLERIES A,B,C,D,G,K

13 Did you read any of the labels or explanations in (PARTICULAR GALLERY)?

Yes ..	1	ASK Q14
No ...	2	SEE Q17

14 Whether people find information interesting can depend on how much they know already. Do you feel that the information in (PARTICULAR GALLERY) is meant

                for someone like yourself 1

RUNNING PROMPT or is it meant for someone who knows less than you do 2

                or for someone who already knows more than you do? 3

                DK 4

15 Do you feel that

                about the right amount of information is provided in (PARTICULAR GALLERY) 1   GO TO Q16

RUNNING PROMPT would you prefer less information. 2

                or would you like there to be more? 3   ASK (a)

IF CODE 3
(a) What other information would you like?

14

16 What about the way the information is presented. Do you feel that the way information is presented in (PARTICULAR GALLERY) could be improved?

         Yes .. 1   ASK (a)

         No 2   SEE Q17

IF YES
(a) How do you feel the way the information is presented could be improved?

         DNA NRM, V&A X   SEE Q17

SCIENCE MUSEUM ONLY
(b) If there were leaflets provided in (PARTICULAR GALLERY) which you could take away and read, would you personally be interested in them?

         Yes .. 1

         No 2

15

Q19 (page 17)

DNA FOR CHILDREN IN UK SCHOOL PARTIES .... X → ASK Q20

MAIN HALL NRM .... Y → ASK (b)

19 TO ALL EXCEPT CHILDREN IN UK SCHOOL PARTIES
Have you been visiting (PARTICULAR GALLERY)

RUNNING PROMPT

because you came across it while walking round the museum ...... 1 → GO TO (b)

or because you had already decided to visit it? ...... 2 → ASK (a)

IF 2
(a) Did you decide to visit (PARTICULAR GALLERY)

RUNNING PROMPT

before you arrived at the museum 1

when you arrived 2

or after you had been in the museum some time? ...... 3 → GO TO Q21

(b) Have you been looking at (PARTICULAR GALLERY)
0 for general interest 1

RUNNING PROMPT

or will this visit to (PARTICULAR GALLERY) help you with your job 2

your studies 3

CODE ALL THAT APPLY

or with a definite hobby? ...... 4 → ASK (i)

OTHER (SPECIFY) ...... 5 → GO TO Q21

ASK SEPARATELY FOR EACH OF 2/3/4 RINGED
(i) How will this visit help you with your job/studies/hobby?
0

17

(page 16)

TO PARENTS VISITING GALLERY WITH OWN CHILDREN AGED 10 OR UNDER

OTHERS DNA .... X → SEE Q19

17 Do you feel that (PARTICULAR GALLERY) is
0 well presented from the point of view of your children ...... 1 → ASK (a)

or do you feel that it wasn't really designed with children like yours in mind? 2

(a) Why do you feel it is [well presented / not well presented] from the point of view of your children?
0

18 Did anything in (PARTICULAR GALLERY) especially attract your children?

Yes .. 1 → ASK (a)

No ... 2 → SEE Q19

IF YES
(a) What especially attracted your children?

(b) May I just check, where are (THINGS) on the plan? NOTE ITEM NUMBERS IN BOX

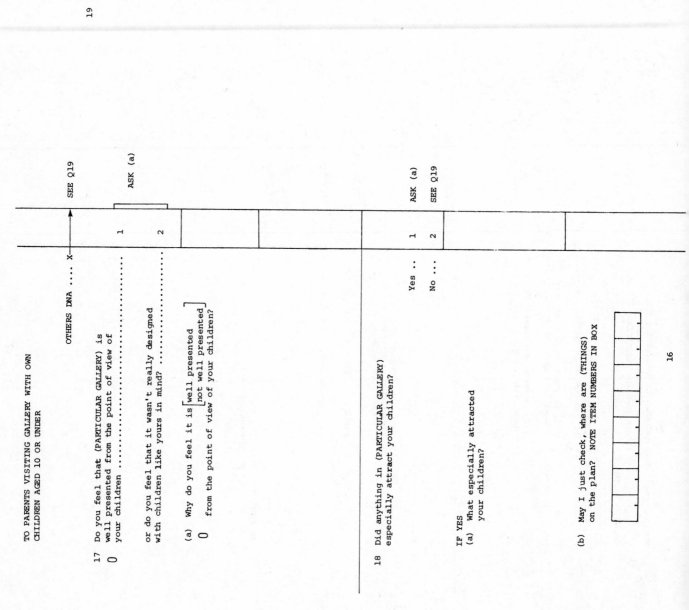

16

TO ALL

21 People visit museums for all kinds of reasons and we have listed a few of them here. (SHOW CARD 1)

0 When you visit a museum - any museum - which reason usually applies to you most?

I go because it is something to do .	1
I go to keep someone else company ..	2
I go for my own interest	3
I go because I have been told to ...	4

22 When you have been round a museum is it important to you to feel that you have

0 learnt something ....

or do you just enjoy looking at things?	1
	2

23 When you look round a room in a museum do you like to think about the different

0 objects individually ....

or do you prefer them to tell an overall story?	1
	2

24 Apart from this museum can you tell me roughly how many museums and art galleries you have visited in the last

0 12 months?

None	1
1	2
2-4	3
5-9	4
10 or more	5
DK	6

CHILDREN IN UK SCHOOL PARTIES ONLY

NRM MAIN HALL ..... X → ASK (a)

20 Were you told by your teacher to visit (PARTICULAR GALLERY) today?

Yes ..	1	ASK (a)
No ...	2	GO TO Q21

IF YES
(a) Is (PARTICULAR GALLERY) connected with what you are learning at school?

0		
Yes ..	1	ASK (i)
No ...	2	ASK (b)

IF YES
(i) How is it connected with what you are doing at school?

0

(b) Were you told by your teachers to do anything in (PARTICULAR GALLERY), such as

RUNNING PROMPT fill in a question and answer sheet 1

make sketches 2

CODE or take notes 3
ALL
THAT or are you just looking around
APPLY (PARTICULAR GALLERY)? 4

OTHER ACTIVITY MENTIONED (SPECIFY) . 5

25 Different museums cover different subjects. Do you personally find some subjects which museums cover more interesting than others?

Yes .. 1   ASK (a)

No ... 2   ASK Q26

IF YES
(a) What subjects are they?

INFORMANT'S BACKGROUND

26 Do you live in the United Kingdom or somewhere else?

UK 1   SEE Q27

Somewhere else .. 2   ASK (a)

IF SOMEWHERE ELSE
(a) What country do you live in?

DNA IRELAND, AUSTRALIA
NEW ZEALAND, USA X   SEE Q27

OTHER RESIDENTS ABROAD
(b) May I just check, what is your first language?

27 Have you completed your full-time education?

DNA 15 OR UNDER X   END

DNA 50 OR OVER .... Y   ASK Q28

Yes .. 1   ASK Q28

No ... 2   ASK Q29

IF FULL-TIME EDUCATION COMPLETED
28 At what age did you finish your full-time education?

DNA FINISHED AGED 16 OR UNDER X   ASK Q30

IF FINISHED AGED 17 OR OVER
(a) What were the main subjects you studied during your last two years of full-time education?

ASK Q30

IF FULL-TIME EDUCATION NOT COMPLETED
29 What are the main subjects you have studied during the last two years?

END

FULL-TIME EDUCATION COMPLETED

30 At present are you

CODE FIRST
THAT APPLIES

working full-time	1
working part-time	2
retired	3
unemployed and looking for work	4
a housewife	5
or not working for some other reason? (SPECIFY) ...	6

ASK Q31 (codes 1–4)

END (codes 5–6)

31 OCCUPATION

IF 1, 2 AT Q30 PRESENT OCCUPATION

IF 3 LAST OCCUPATION PRIOR TO RETIREMENT

IF 4 LAST OCCUPATION

NOTE : IF TEACHER/LECTURER
PROBE FOR SUBJECTS TEACHES
AND TYPE OF SCHOOL/COLLEGE
TEACHES AT

INTERVIEWER
CHECKS

S/E	1
Employee	2
Manager	1
Foreman/Sup	2
Other	3

Social Survey Division
Office of Population Censuses and Surveys
St Catherines House
10 Kingsway
London WC2B 6JP

W849 OPCS 9/79

22

Index

153

Printed in the UK for Her Majesty's Stationery Office
Dd737143 C9 8/84 10170 (1090)